EX LIBRIS
SOUTH ORANGE
PUBLIC LIBRARY

*Environmental Issues*

# WILDLIFE PROTECTION

*Environmental Issues*

AIR QUALITY

CLIMATE CHANGE

CONSERVATION

ENVIRONMENTAL POLICY

WATER POLLUTION

WILDLIFE PROTECTION

Environmental Issues

# WILDLIFE PROTECTION

Yael Calhoun
Series Editor

Foreword by David Seideman,
Editor-in-Chief, *Audubon* Magazine

Philadelphia

CHELSEA HOUSE PUBLISHERS
VP, NEW PRODUCT DEVELOPMENT Sally Cheney
DIRECTOR OF PRODUCTION Kim Shinners
CREATIVE MANAGER Takeshi Takahashi
MANUFACTURING MANAGER Diann Grasse

Staff for WILDLIFE PROTECTION
EXECUTIVE EDITOR Tara Koellhoffer
EDITORIAL ASSISTANT Kuorkor Dzani
PRODUCTION EDITOR Noelle Nardone
PHOTO EDITOR Sarah Bloom
SERIES AND COVER DESIGNER Keith Trego
LAYOUT 21st Century Publishing and Communications, Inc.


www.chelseahouse.com

A Haights Cross Communications Company ®

First Printing

9 8 7 6 5 4 3 2 1

ibrary of Congress Cataloging-in-Publication Data

Wildlife protection/edited by Yael Calhoun; foreword by David Seideman.
p. cm.—(Environmental issues)
Includes bibliographical references and index.
ISBN 0-7910-8204-0
1. Wildlife conservation. I. Calhoun, Yael. II. Series.
QL82.W56 2005
333.95'416—dc22

2004029000

All links and web addresses were checked and verified to be correct at the time of publication. Because of the dynamic nature of the web, some addresses and links may have changed since publication and may no longer be valid.

# Contents Overview

Foreword by David Seideman, Editor-in-Chief, *Audubon* Magazine x
Introduction: "Why Should We Care?" xvi

Section A:
**Wildlife Protection Issues and Challenges** 1

Section B:
**Endangered Species** 21

Section C:
**Reserves for Wildlife Protection** 61

Section D:
**Other Programs for Wildlife Protection** 87

Bibliography 112
Further Reading 114
Index 115

# Detailed Table of Contents

Foreword by David Seideman, Editor-in-Chief, *Audubon* Magazine x

Introduction: "Why Should We Care?" xvi

Section A:

**Wildlife Protection Issues and Challenges** 1

**What Is Biodiversity?** 2

**Biodiversity** 3
by Boyce Rensberger

**Why Are Biologists Concerned About the Decline of Frog Populations?** 9

**Amphibian Conservation** 10
by Raymond D. Semlitsch, ed.

**What Do Declining Bird Populations Tell Us About Our Environment?** 13

**Winged Messengers: The Decline of Birds** 14
by Howard Youth

Section B.

**Endangered Species** 21

**Is the Concern Over Whale Populations a New Issue?** 22

**Whaling** 23
from the World Wildlife Fund

**What Are Countries Around the World Doing to Protect Endangered Species?** 30

**CITES** 30
from the World Wildlife Fund

**How Are Endangered Species Protected in the United States?** 33

**Endangered Species Bulletin No. 4.** 34
from the U.S. Fish and Wildlife Service Report

What Dangers Do Wolves Still Face in Our Country? 40

America's Wolves Threatened Again 41
by Nina Fascione

Are Protection Efforts Working for Endangered Species Around the World? 48

Flagship Species Factsheets 49
from the World Wildlife Fund

Section C:
Reserves For Wildlife Protection 61

How Do Conservation Organizations Work With Local People, Businesses, and Governments to Protect Land and the Plants and Animals That Live There? 62

Conserving Nature: Partnering With People—World Wildlife Fund's Global Work on Protected Area Networks 63
from the World Wildlife Fund

What Programs Serve to Protect Our Important Ocean Habitats? 72

Marine Reserves: A Tool for Ecosystem Management and Conservation 73
from the Pew Oceans Commission

Is Our National Wildlife Refuge System Working? 78

The Art of Helping Wildlife 79
by T. Edward Nickens

Section D:
**Other Programs for Wildlife Protection** 87

**What Programs Are Protecting the Endangered Orangutans?** 88

**Summary of Orangutan Surveys Conducted in Berau District, East Kalimantan** 89
from The Nature Conservancy

**The Nature Conservancy Finds Population of Wild Orangutans** 91
from The Nature Conservancy

**The Nature Conservancy Uses Computerized Bar Codes to Stop Illegal Logging in Indonesia** 94
from The Nature Conservancy

**Do Programs Developed by Conservation Organizations to Protect Habitat and Endangered Animals Work?** 97

**The Hyacinth Macaw Makes a Comeback** 98
by Harold Palo

**A Royal Return** 103
by Karen Coates

**Hunting for Conservation** 108
from the World Wildlife Fund and Wildlife Conservation Society

Bibliography 112

Further Reading 114

Index 115

# Foreword

by David Seideman, Editor-in-Chief, *Audubon* Magazine

For anyone contemplating the Earth's fate, there's probably no more instructive case study than the Florida Everglades. When European explorers first arrived there in the mid-1800s, they discovered a lush, tropical wilderness with dense sawgrass, marshes, mangrove forests, lakes, and tree islands. By the early $20^{th}$ century, developers and politicians had begun building a series of canals and dikes to siphon off the region's water. They succeeded in creating an agricultural and real estate boom, and to some degree, they offset floods and droughts. But the ecological cost was exorbitant. Today, half of the Everglades' wetlands have been lost, its water is polluted by runoff from farms, and much of its wildlife, including Florida panthers and many wading birds such as wood storks, are hanging on by a thread.

Yet there has been a renewed sense of hope in the Everglades since 2001, when the state of Florida and the federal government approved a comprehensive $7.8 billion restoration plan, the biggest recovery of its kind in history. During the next four decades, ecologists and engineers will work to undo years of ecological damage by redirecting water back into the Everglades' dried-up marshes. "The Everglades are a test," says Joe Podger, an environmentalist. "If we pass, we get to keep the planet."

In fact, as this comprehensive series on environmental issues shows, humankind faces a host of tests that will determine whether we get to keep the planet. The world's crises—air and water pollution, the extinction of species, and climate change—are worsening by the day. The solutions—and there are many practical ones—all demand an extreme sense of urgency. E. O. Wilson, the noted Harvard zoologist, contends that "the world environment is changing so fast that there is a window of opportunity that will close in as little time as the next two or three decades." While Wilson's main concern is the rapid loss of biodiversity, he could have just as easily been discussing climate change or wetlands destruction.

The Earth is suffering the most massive extinction of species since the die-off of dinosaurs 65 million years ago. "If we continue at the current rate of deforestation and destruction of major ecosystems like rain forests and coral reefs, where most of the biodiversity is concentrated," Wilson says, "we will surely lose more than half of all the species of plants and animals on Earth by the end of the 21st century."

Many conservationists still mourn the loss of the passenger pigeon, which, as recently as the late 1800s, flew in miles-long flocks so dense they blocked the sun, turning noontime into nighttime. By 1914, target shooters and market hunters had reduced the species to a single individual, Martha, who lived at the Cincinnati Zoo until, as Peter Matthiessen wrote in *Wildlife in America,* "she blinked for the last time." Despite U.S. laws in place to avert other species from going the way of the passenger pigeon, the latest news is still alarming. In its 2004 State of the Birds report, Audubon noted that 70% of grassland bird species and 36% of shrubland bird species are suffering significant declines. Like the proverbial canary in the coalmine, birds serve as indicators, sounding the alarm about impending threats to environmental and human health.

Besides being an unmitigated moral tragedy, the disappearance of species has profound practical implications. Ninety percent of the world's food production now comes from about a dozen species of plants and eight species of livestock. Geneticists rely on wild populations to replenish varieties of domestic corn, wheat, and other crops, and to boost yields and resistance to disease. "Nature is a natural pharmacopoeia, and new drugs and medicines are being discovered in the wild all the time," wrote Niles Eldredge of the American Museum of Natural History, a noted author on the subject of extinction. "Aspirin comes from the bark of willow trees. Penicillin comes from a mold, a type of fungus." Furthermore, having a wide array of plants and animals improves a region's capacity to cleanse water, enrich soil, maintain stable climates, and produce the oxygen we breathe.

Today, the quality of the air we breathe and the water we drink does not augur well for our future health and well-being. Many people assume that the passage of the Clean Air Act in 1970 ushered in a new age. But the American Lung Association reports that 159 million Americans—55% of the population—are exposed to "unhealthy levels of air pollution." Meanwhile, the American Heart Association warns of a direct link between exposure to air pollution and heart disease and strokes. While it's true that U.S. waters are cleaner than they were three decades ago, data from the Environmental Protection Agency (EPA) shows that almost half of U.S. coastal waters fail to meet water-quality standards because they cannot support fishing or swimming. Each year, contaminated tap water makes as many as 7 million Americans sick. The chief cause is "non-point pollution," runoff that includes fertilizers and pesticides from farms and backyards as well as oil and chemical spills. On a global level, more than a billion people lack access to clean water; according to the United Nations, five times that number die each year from malaria and other illnesses associated with unsafe water.

Of all the Earth's critical environmental problems, one trumps the rest: climate change. Carol Browner, the EPA's chief from 1993 through 2001 (the longest term in the agency's history), calls climate change "the greatest environmental health problem the world has ever seen." Industry and people are spewing carbon dioxide from smokestacks and the tailpipes of their cars into the atmosphere, where a buildup of gases, acting like the glass in a greenhouse, traps the sun's heat. The 1990s was the warmest decade in more than a century, and 1998 saw the highest global temperatures ever. In an article about global climate change in the December 2003 issue of *Audubon*, David Malakoff wrote, "Among the possible consequences: rising sea levels that cause coastal communities to sink beneath the waves like a modern Atlantis, crop failures of biblical proportions, and once-rare killer storms that start to appear with alarming regularity."

Yet for all the doom and gloom, scientists and environmentalists hold out hope. When Russia recently ratified the Kyoto Protocol, it meant that virtually all of the world's industrialized nations—the United States, which has refused to sign, is a notable exception—have committed to cutting greenhouse gases. As Kyoto and other international agreements go into effect, a market is developing for cap-and-trade systems for carbon dioxide. In this country, two dozen big corporations, including British Petroleum, are cutting emissions. At least 28 American states have adopted their own policies. California, for example, has passed global warming legislation aimed at curbing emissions from new cars. Governor Arnold Schwarzenegger has also backed regulations requiring automakers to slash the amount of greenhouse gases they cause by up to 30% by 2016, setting a precedent for other states.

As Washington pushes a business-friendly agenda, states are filling in the policy vacuum in other areas, as well. California and New York are developing laws to preserve wetlands, which filter pollutants, prevent floods, and provide habitat for endangered wildlife.

By taking matters into their own hands, states and foreign countries will ultimately force Washington's. What industry especially abhors is a crazy quilt of varying rules. After all, it makes little sense for a company to invest a billion dollars in a power plant only to find out later that it has to spend even more to comply with another state's stricter emissions standards. Ford chairman and chief executive William Ford has lashed out at the states' "patchwork" approach because he and "other manufacturers will have a hard time responding." Further, he wrote in a letter to his company's top managers, "the prospect of 50 different requirements in 50 different states would be nothing short of chaos." The type of fears Ford expresses are precisely the reason federal laws protecting clean air and water came into being.

Governments must take the lead, but ecologically conscious consumers wield enormous influence, too. Over the past four decades, the annual use of pesticides has more than doubled, from 215 million pounds to 511 million pounds. Each year, these poisons cause $10 billion worth of damage to the environment and kill 72 million birds. The good news is that the demand for organic products is revolutionizing agriculture, in part by creating a market for natural alternatives for pest control. Some industry experts predict that by 2007 the organic industry will almost quadruple, to more than $30 billion.

E. O. Wilson touts "shade-grown" coffee as one of many "personal habitats that, if moderated only in this country, could contribute significantly to saving endangered species." In the mountains of Mexico and Central America, coffee grown beneath a dense forest canopy rather than in cleared fields helps provide refuge for dozens of wintering North American migratory bird species, from western tanagers to Baltimore orioles.

With conservation such a huge part of Americans' daily routine, recycling has become as ingrained a civic duty as obeying traffic lights. Californians, for their part, have cut their energy consumption by 10% each year since the state's 2001 energy crisis. "Poll after poll shows that about two-thirds of the American public—Democrat and Republican, urban and rural—consider environmental progress crucial," writes Carl Pope, director of the Sierra Club, in his recent book, *Strategic Ignorance.* "Clean air, clean water, wilderness preservation—these are such bedrock values that many polling respondents find it hard to believe that any politician would oppose them."

Terrorism and the economy clearly dwarfed all other issues in the 2004 presidential election. Even so, voters approved 120 out of 161 state and local conservation funding measures nationwide, worth a total of $3.25 billion. Anti-environment votes in the U.S. Congress and proposals floated by the like-minded Bush administration should not obscure the salient fact that so far there have been no changes to the major environmental laws. The potential for political fallout is too great.

The United States' legacy of preserving its natural heritage is the envy of the world. Our national park system alone draws more than 300 million visitors each year. Less well known is the 103-year-old national wildlife refuge system you'll learn about in this series. Its unique mission is to safeguard the nation's wild animals and plants on 540 refuges, protecting 700 species of birds and an equal number of other vertebrates; 282 of these species are either threatened or endangered. One of the many species particularly dependent on the invaluable habitat refuges afford is the bald eagle. Such safe havens, combined with the banning of the insecticide DDT and enforcement of the

Endangered Species Act, have led to the bald eagle's remarkable recovery, from a low of 500 breeding pairs in 1963 to 7,600 today. In fact, this bird, the national symbol of the United States, is about be removed from the endangered species list and downgraded to a less threatened status under the CITES, the Convention on International Trade in Endangered Species.

This vital treaty, upheld by the United States and 165 other participating nations (and detailed in this series), underscores the worldwide will to safeguard much of the Earth's magnificent wildlife. Since going into effect in 1975, CITES has helped enact plans to save tigers, chimpanzees, and African elephants. These species and many others continue to face dire threats from everything from poaching to deforestation. At the same time, political progress is still being made. Organizations like the World Wildlife Fund work tirelessly to save these species from extinction because so many millions of people care. China, for example, the most populous nation on Earth, is so concerned about its giant pandas that it has implemented an ambitious captive breeding program. That program's success, along with government measures prohibiting logging throughout the panda's range, may actually enable the remaining population of 1,600 pandas to hold its own—and perhaps grow. "For the People's Republic of China, pressure intensified as its internationally popular icon edged closer to extinction," wrote Gerry Ellis in a recent issue of *National Wildlife.* "The giant panda was not only a poster child for endangered species, it was a symbol of our willingness to ensure nature's place on Earth."

Whether people take a spiritual path to conservation or a pragmatic one, they ultimately arrive at the same destination. The sight of a bald eagle soaring across the horizon reassures us about nature's resilience, even as the clean air and water we both need to survive becomes less of a certainty. "The conservation of our natural resources and their proper use constitute the fundamental problem which underlies almost every other problem of our national life," President Theodore Roosevelt told Congress at the dawn of the conservation movement a century ago. His words ring truer today than ever.

# Introduction: "Why Should We Care?"

Our nation's air and water are cleaner today than they were 30 years ago. After a century of filling and destroying over half of our wetlands, we now protect many of them. But the Earth is getting warmer, habitats are being lost to development and logging, and humans are using more water than ever before. Increased use of water can leave rivers, lakes, and wetlands without enough water to support the native plant and animal life. Such changes are causing plants and animals to go extinct at an increased rate. It is no longer a question of losing just the dodo birds or the passenger pigeons, argues David Quammen, author of *Song of the Dodo*: "Within a few decades, if present trends continue, we'll be losing *a lot* of everything."[1]

In the 1980s, E. O. Wilson, a Harvard biologist and Pulitzer Prize–winning author, helped bring the term *biodiversity* into public discussions about conservation. *Biodiversity*, short for "biological diversity," refers to the levels of organization for living things. Living organisms are divided and categorized into ecosystems (such as rain forests or oceans), by species (such as mountain gorillas), and by genetics (the genes responsible for inherited traits).

Wilson has predicted that if we continue to destroy habitats and pollute the Earth at the current rate, in 50 years, we could lose 30 to 50% of the planet's species to extinction. In his 1992 book, *The Diversity of Life*, Wilson asks: "Why should we care?"[2] His long list of answers to this question includes: the potential loss of vast amounts of scientific information that would enable the development of new crops, products, and medicines and the potential loss of the vast economic and environmental benefits of healthy ecosystems. He argues that since we have only a vague idea (even with our advanced scientific methods) of how ecosystems really work, it would be "reckless" to suppose that destroying species indefinitely will not threaten us all in ways we may not even understand.

## THE BOOKS IN THE SERIES

In looking at environmental issues, it quickly becomes clear that, as naturalist John Muir once said, "When we try to pick out anything by itself, we find it hitched to everything else in the Universe."[3] For example, air pollution in one state or in one country can affect not only air quality in another place, but also land and water quality. Soil particles from degraded African lands can blow across the ocean and cause damage to far-off coral reefs.

The six books in this series address a variety of environmental issues: conservation, wildlife protection, water pollution, air quality, climate change, and environmental policy. None of these can be viewed as a separate issue. Air quality impacts climate change, wildlife, and water quality. Conservation initiatives directly affect water and air quality, climate change, and wildlife protection. Endangered species are touched by each of these issues. And finally, environmental policy issues serve as important tools in addressing all the other environmental problems that face us.

You can use the burning of coal as an example to look at how a single activity directly "hitches" to a variety of environmental issues. Humans have been burning coal as a fuel for hundreds of years. The mining of coal can leave the land stripped of vegetation, which erodes the soil. Soil erosion contributes to particulates in the air and water quality problems. Mining coal can also leave piles of acidic tailings that degrade habitats and pollute water. Burning any fossil fuel—coal, gas, or oil—releases large amounts of carbon dioxide into the atmosphere. Carbon dioxide is considered a major "greenhouse gas" that contributes to global warming—the gradual increase in the Earth's temperature over time. In addition, coal burning adds sulfur dioxide to the air, which contributes to the formation of acid rain—precipitation that is abnormally acidic. This acid rain can kill forests and leave lakes too acidic to support life. Technology continues to present ways to minimize the pollution that results from extracting and burning fossil fuels. Clean air and climate change policies guide states and industries toward implementing various strategies and technologies for a cleaner coal industry.

Each of the six books in this series—ENVIRONMENTAL ISSUES—introduces the significant points that relate to the specific topic and explains its relationship to other environmental concerns.

### Book One: *Air Quality*

Problems of air pollution can be traced back to the time when humans first started to burn coal. *Air Quality* looks at today's challenges in fighting to keep our air clean and safe. The book includes discussions of air pollution sources—car and truck emissions, diesel engines, and many industries. It also discusses their effects on our health and the environment.

The Environmental Protection Agency (EPA) has reported that more than 150 million Americans live in areas that have unhealthy levels of some type of air pollution.[4] Today, more than 20 million Americans, over 6 million of whom are children, suffer from asthma believed to be triggered by pollutants in the air.[5]

In 1970, Congress passed the Clean Air Act, putting in place an ambitious set of regulations to address air pollution concerns. The EPA has identified and set standards for six common air pollutants: ground-level ozone, nitrogen oxides, particulate matter, sulfur dioxide, carbon monoxide, and lead.

The EPA has also been developing the Clean Air Rules of 2004, national standards aimed at improving the country's air quality by specifically addressing the many sources of contaminants. However, many conservation organizations and even some states have concerns over what appears to be an attempt to weaken different sections of the 1990 version of the Clean Air Act. The government's environmental protection efforts take on increasing importance because air pollution degrades land and water, contributes to global warming, and affects the health of plants and animals, including humans.

### Book Two: *Climate Change*

Part of science is observing patterns, and scientists have observed a global rise in temperature. *Climate Change* discusses the sources and effects of global warming. Scientists attribute this accelerated change to human activities such as the burning of fossil fuels that emit greenhouse gases (GHG).[6] Since the 1700s, we have been cutting down the trees that help remove carbon dioxide from the atmosphere, and have increased the amount of coal, gas, and oil we burn, all of which add carbon dioxide to the atmosphere. Science tells us that these human activities have caused greenhouse gases—carbon dioxide ($CO_2$), methane ($CH_4$), nitrous oxide ($N_2O$), hydrofluorocarbons (HFCs), perfluorocarbons (PFCs), and sulfur hexafluoride ($SF_6$)—to accumulate in the atmosphere.[7]

If the warming patterns continue, scientists warn of more negative environmental changes. The effects of climate change, or global warming, can be seen all over the world. Thousands of scientists are predicting rising sea levels, disturbances in patterns of rainfall and regional weather, and changes in ranges and reproductive cycles of plants and animals. Climate change is already having some effects on certain plant and animal species.[8]

Many countries and some American states are already working together and with industries to reduce the emissions of greenhouse gases. Climate change is an issue that clearly fits noted scientist Rene Dubois's advice: "Think globally, act locally."

### Book Three: *Conservation*

*Conservation* considers the issues that affect our world's vast array of living creatures and the land, water, and air they need to survive.

One of the first people in the United States to put the political spotlight on conservation ideas was President Theodore Roosevelt. In the early 1900s, he formulated policies and created programs that addressed his belief that: "The nation behaves well if it treats the natural resources as assets which it must turn over to the next generation increased, and not impaired, in value."[9] In the 1960s, biologist Rachel Carson's book, *Silent Spring*, brought conservation issues into the public eye. People began to see that polluted land, water, and air affected their health. The 1970s brought the creation of the United States Environmental Protection Agency (EPA) and passage of many federal and state rules and regulations to protect the quality of our environment and our health.

Some 80 years after Theodore Roosevelt established the first National Wildlife Refuge in 1903, Harvard biologist

E. O. Wilson brought public awareness of conservation issues to a new level. He warned:

> . . . the worst thing that will probably happen—in fact is already well underway—is not energy depletion, economic collapse, conventional war, or even the expansion of totalitarian governments. As terrible as these catastrophes would be for us, they can be repaired within a few generations. The one process now ongoing that will take million of years to correct is the loss of genetic species diversity by the destruction of natural habitats. This is the folly our descendants are least likely to forgive us.[10]

To heed Wilson's warning means we must strive to protect species-rich habitats, or "hotspots," such as tropical rain forests and coral reefs. It means dealing with conservation concerns like soil erosion and pollution of fresh water and of the oceans. It means protecting sea and land habitats from the overexploitation of resources. And it means getting people involved on all levels—from national and international government agencies, to private conservation organizations, to the individual person who recycles or volunteers to listen for the sounds of frogs in the spring.

### Book Four: *Environmental Policy*

One approach to solving environmental problems is to develop regulations and standards of safety. Just as there are rules for living in a community or for driving on a road, there are environmental regulations and policies that work toward protecting our health and our lands. *Environmental Policy* discusses the regulations and programs that have been crafted to address environmental issues at all levels—global, national, state, and local.

Today, as our resources become increasingly limited, we witness heated debates about how to use our public lands and how to protect the quality of our air and water. Should we allow drilling in the Arctic National Wildlife Refuge? Should we protect more marine areas? Should we more closely regulate the emissions of vehicles, ships, and industries? These policy issues, and many more, continue to make news on a daily basis.

In addition, environmental policy has taken a place on the international front. Hundreds of countries are working together in a variety of ways to address such issues as global warming, air pollution, water pollution and supply, land preservation, and the protection of endangered species. One question the United States continues to debate is whether to sign the 1997 Kyoto Protocol, the international agreement designed to decrease the emissions of greenhouse gases.

Many of the policy tools for protecting our environment are already in place. It remains a question how they will be used—and whether they will be put into action in time to save our natural resources and ourselves.

### Book Five: *Water Pollution*

Pollution can affect water everywhere. Pollution in lakes and rivers is easily seen. But water that is out of our plain view can also be polluted with substances such as toxic chemicals, fertilizers, pesticides, oils, and gasoline. *Water Pollution* considers issues of concern to our surface waters, our groundwater, and our oceans.

In the early 1970s, about three-quarters of the water in the United States was considered unsafe for swimming and fishing. When Lake Erie was declared "dead" from pollution and a river feeding it actually caught on fire, people decided that the national government had to take a stronger role in protecting our resources. In 1972, Congress passed the Clean Water Act, a law whose objective "is to restore and maintain the chemical, physical, and biological integrity of the Nation's waters."[11] Today, over 30 years later, many lakes and rivers have been restored to health. Still, an estimated 40% of our waters are still unsafe to swim in or fish.

Less than 1% of the available water on the planet is fresh water. As the world's population grows, our demand for drinking and irrigation water increases. Therefore, the quantity of available water has become a major global issue. As Sandra Postel, a leading authority on international freshwater issues, says, "Water scarcity is now the single biggest threat to global food production."[12] Because there are many competing demands for water, including the needs of habitats, water pollution continues to become an even more serious problem each year.

### Book Six: *Wildlife Protection*

For many years, the word *wildlife* meant only the animals that people hunted for food or for sport. It was not until 1986 that the Oxford English Dictionary defined *wildlife* as "the native fauna and flora of a particular region."[13] *Wildlife Protection* looks at overexploitation—for example, overfishing or collecting plants and animals for illegal trade—and habitat loss. Habitat loss can be the result of development, logging, pollution, water diverted for human use, air pollution, and climate change.

Also discussed are various approaches to wildlife protection. Since protection of wildlife is an issue of global concern, it is addressed here on international as well as on national and local levels. Topics include voluntary international organizations such as the International Whaling Commission and the CITES agreements on trade in endangered species. In the United States, the Endangered Species Act provides legal protection for more than 1,200 different plant and animal species. Another approach to wildlife protection includes developing partnerships among conservation organizations, governments, and local people to foster economic incentives to protect wildlife.

## CONSERVATION IN THE UNITED STATES

Those who first lived on this land, the Native American peoples, believed in general that land was held in common, not to be individually owned, fenced, or tamed. The white settlers from Europe had very different views of land. Some believed the New World was a Garden of Eden. It was a land of opportunity for them, but it was also a land to be controlled and subdued. Ideas on how to treat the land often followed those of European thinkers like John Locke, who believed that "Land that is left wholly to nature is called, as indeed it is, waste."[14]

The 1800s brought another way of approaching the land. Thinkers such as Ralph Waldo Emerson, John Muir, and Henry David Thoreau celebrated our human connection with nature. By the end of the 1800s, some scientists and policymakers were noticing the damage humans have caused to the land. Leading public officials preached stewardship and wise use of our country's resources. In 1873, Yellowstone National Park was set up. In 1903, the first National Wildlife Refuge was established.

However, most of the government practices until the middle of the 20th century favored unregulated development and use of the land's resources. Forests were clear cut, rivers were dammed, wetlands were filled to create farmland, and factories were allowed to dump their untreated waste into rivers and lakes.

In 1949, a forester and ecologist named Aldo Leopold revived the concept of preserving land for its own sake. But there was now a biological, or scientific, reason for conservation, not just a spiritual one. Leopold declared: "All ethics rest upon a single premise: that the individual is a member of a community of interdependent parts. . . . A thing is right when it tends to preserve the integrity and stability and beauty of the biotic community. It is wrong when it tends otherwise."[15]

The fiery vision of these conservationists helped shape a more far-reaching movement that began in the 1960s. Many credit Rachel Carson's eloquent and accessible writings, such as her 1962 book *Silent Spring*, with bringing environmental issues into people's everyday language. When the Cuyahoga River in Ohio caught fire in 1969 because it was so polluted, it captured the public attention. Conservation was no longer just about protecting land that many people would never even see, it was about protecting human health. The condition of the environment had become personal.

In response to the public outcry about water and air pollution, the 1970s saw the establishment of the EPA. Important legislation to protect the air and water was passed. National standards for a cleaner environment were set and programs were established to help achieve the ambitious goals. Conservation organizations grew from what had started as exclusive white men's hunting clubs to interest groups with a broad membership base. People came together to demand changes that would afford more protection to the environment and to their health.

Since the 1960s, some presidential administrations have sought to strengthen environmental protection and to protect more land and national treasures. For example, in 1980, President Jimmy Carter signed an act that doubled the amount of protected land in Alaska and renamed it the Arctic National Wildlife Refuge. Other administrations, like those of President Ronald Reagan, sought to dismantle many earlier environmental protection initiatives.

The environmental movement, or environmentalism, is not one single, homogeneous cause. The agencies, individuals, and organizations that work toward protecting the environment vary as widely as the habitats and places they seek to protect. There are individuals who begin grass-roots efforts—people like Lois Marie Gibbs, a former resident of the polluted area of Love Canal, New York, who founded the Center for Health, Environment and Justice. There are conservation organizations, like The Nature Conservancy, the World Wildlife Fund (WWF), and Conservation International, that sponsor programs to preserve and protect habitats. There are groups that specialize in monitoring public policy and legislation—for example, the Natural Resources Defense Council and Environmental Defense. In addition, there are organizations like the Audubon Society and the National Wildlife Federation whose focus is on public education about environmental issues. Perhaps from this diversity, just like there exists in a healthy ecosystem, will come the strength and vision environmentalism needs to deal with the continuing issues of the 21st century.

## INTERNATIONAL CONSERVATION EFFORTS

In his book *Biodiversity*, E. O. Wilson cautions that biological diversity must be taken seriously as a global resource for three reasons. First, human population growth is accelerating the degrading of the environment, especially in tropical countries. Second, science continues to discover new uses for biological diversity—uses that can benefit human health and protect the environment. And third, much biodiversity is being lost through extinction, much of it in the tropics. As Wilson states, "We must hurry to acquire the knowledge on which a wise policy of conservation and development can be based for centuries to come."[16]

People organize themselves within boundaries and borders. But oceans, rivers, air, and wildlife do not follow such rules. Pollution or overfishing in one part of an ocean can easily degrade the quality of another country's resources. If one country diverts a river, it can destroy another country's wetlands or water resources. When Wilson cautions us that we must hurry to develop a wise conservation policy, he means a policy that will protect resources all over the world.

To accomplish this will require countries to work together on critical global issues: preserving biodiversity, reducing global warming, decreasing air pollution, and protecting the oceans. There are many important international efforts already going on to protect the resources of our planet. Some efforts are regulatory, while others are being pursued by nongovernmental organizations or private conservation groups.

Countries volunteering to cooperate to protect resources is not a new idea. In 1946, a group of countries established the International Whaling Commission (IWC). They recognized that unregulated whaling around the world had led to severe declines in the world's whale populations. In 1986, the IWC declared a moratorium on whaling, which is still in effect, until the populations have recovered.[17] Another example of international cooperation occurred in 1987 when various countries signed the Montreal Protocol to reduce the emissions of ozone-depleting gases. It has been a huge success, and perhaps has served as a model for other international efforts, like the 1997 Kyoto Protocol, to limit emissions of greenhouse gases.

Yet another example of international environmental cooperation is the CITES agreement (the Convention on International Trade in Endangered Species of Wild Fauna and Flora), a legally binding agreement to ensure that the international trade of plants and animals does not threaten the species' survival. CITES went into force in 1975 after 80 countries agreed to the terms. Today, it has grown to include more than 160 countries. This make CITES among the largest conservation agreements in existence.[18]

Another show of international conservation efforts are governments developing economic incentives for local conservation. For example, in 1996, the International Monetary Fund (IMF) and the World Wildlife Fund (WWF) established a program to relieve poor countries of debt. More than 40 countries have benefited by agreeing to direct some of their savings toward environmental programs in the "Debt-for-Nature" swap programs.[19]

It is worth our time to consider the thoughts of two American conservationists and what role we, as individuals, can play in conserving and protecting our world. E. O. Wilson has told us that "Biological Diversity—'biodiversity' in the new parlance—is the key to the maintenance of the world as we know it."[20] Aldo Leopold, the forester who gave Americans the idea of creating a "land ethic," wrote in 1949 that: "Having to squeeze the last drop of utility out of the land has the same desperate finality as having to chop up the furniture to keep warm."[21] All of us have the ability to take part in the struggle to protect our environment and to save our endangered Earth.

## ENDNOTES

1 Quammen, David. *Song of the Dodo.* New York: Scribner, 1996, p. 607.

2 Wilson, E. O. *Diversity of Life.* Cambridge, MA: Harvard University Press, 1992, p. 346.

3 Muir, John. *My First Summer in the Sierra.* San Francisco: Sierra Club Books, 1988, p. 110.

4 Press Release. *EPA Newsroom: EPA Issues Designations on Ozone Health Standards.* April 15, 2004. Available online at *http://www.epa.gov/newsroom/.*

5 The Environmental Protection Agency. EPA Newsroom. *May is Allergy Awareness Month.* May 2004. Available online at *http://www.epa.gov/newsroom/allergy_month.htm.*

6 Intergovernmental Panel on Climate Change (IPCC). Third Annual Report, 2001.

7 Turco, Richard P. *Earth Under Siege: From Air Pollution to Global Change.* New York: Oxford University Press, 2002, p. 387.

8 Intergovernmental Panel on Climate Change. *Technical Report V: Climate Change and Biodiversity.* 2002. Full report available online at *http://www.ipcc.ch/pub/tpbiodiv.pdf.*

9 "Roosevelt Quotes." American Museum of Natural History. Available online at *http://www.amnh.org/common/faq/quotes.html.*

10 Wilson, E. O. *Biophilia.* Cambridge, MA: Harvard University Press, 1986, pp. 10–11.

11 Federal Water Pollution Control Act. As amended November 27, 2002. Section 101 (a).

12 Postel, Sandra. *Pillars of Sand.* New York: W. W. Norton & Company, Inc., 1999. p. 6.

13 Hunter, Malcolm L. *Wildlife, Forests, and Forestry: Principles of Managing Forest for Biological Diversity.* Englewood Cliffs, NJ: Prentice-Hall, 1990, p. 4.

14 Dowie, Mark. *Losing Ground: American Environmentalism at the Close of the Twentieth Century.* Cambridge, MA: MIT Press, 1995, p. 113.

15 Leopold, Aldo. *A Sand County Almanac.* New York: Oxford University Press, 1949.

16 Wilson, E. O., ed. *Biodiversity.* Washington, D.C.: National Academies Press, 1988, p. 3.

17 International Whaling Commission Information 2004. Available online at *http://www.iwcoffice.org/commission/iwcmain.htm.*

18 *Discover CITES: What is CITES?* Fact sheet 2004. Available online at *http://www.cites.org/eng/disc/what.shtml.*

19 *Madagascar's Experience with Swapping Debt for the Environment.* World Wildlife Fund Report, 2003. Available online at *http://www.conservationfinance.org/WPC/WPC_documents/Apps_11_Moye_Paddack_v2.pdf.*

20 Wilson, *Diversity of Life*, p. 15.

21 Leopold.

# What Is Biodiversity?

E. O. Wilson, a Harvard biologist, helped bring the concept of biological diversity into the conservation movement. As he explains in the following interview, he realized he had to "cross over and get involved because the biodiversity was just disappearing too fast." Wilson has been called the "father" of the term *biodiversity*. Biodiversity, short for "biological diversity," is organized into three levels: ecosystems (for example, rain forests or deserts), species (the organisms in an ecosystem—humans or red maple trees are examples), and the variety of genes that make up the organisms (genes are responsible for heredity).[1] Wilson has predicted that if humans keep altering the Earth, that we could lose one-fifth, or 20%, or more of our plant and animal biodiversity by the year 2020.[2]

In the following interview, Wilson notes the shifts in the focus of conservation organizations from protecting particular species to protecting the entire ecosystem in which they live. Although organizations play key roles in preserving biodiversity, there is also an important role for individuals. For example, you can help by making informed decisions about the kinds of products you buy. Wilson says that choosing shade-grown coffee, not eating imported beef, and not buying products made from endangered species are all ways to help protect biodiversity.

Wilson believes that the fate of the world's flora and fauna depends on "a combination of science, education, and ethics." He cautions that the window of opportunity for protecting our biodiversity could be as little as two or three decades.

Perhaps we would be wise to consider the words of John Sawhill, president of The Nature Conservancy from 1990 to 2000, words that E. O. Wilson himself put at the beginning of his 2002 book, *The Future of Life*: "In the end, our society will be defined no only by what we create, but by what we refuse to destroy."[3]

—The Editor

1. Wilson, E. O. *The Future of Life*. New York: Random House, 2002, pp. 10–11.
2. Wilson, E. O. *The Diversity of Life.* Cambridge, MA: Harvard University Press, 1992, p. 246.
3. Wilson, *The Future of Life*, pp. 10–11.

## Biodiversity

by Boyce Rensberger

If, a century from now, earthlings gaze upon their planet and wonder at its extraordinary diversity of life, they are likely to hold in reverence the name of one man: Edward O. Wilson. Indeed, many today who fear the rapid destruction of biological diversity already regard Wilson as the nearest thing to a savior that science has produced.

Earth's biodiversity has not been saved—not yet. Uncounted species are endangered, flickering out of existence in large numbers before they can even be named. But the man who has done more than almost anyone else to focus society's attention on the value of biological diversity, and the threats to it, says there is reason for hope. Though he would be the last to say so, Wilson's campaign on behalf of the living world has helped make the conservation movement's goals attainable.

No starry-eyed radical, the Harvard University zoologist helped establish the concept of biological diversity, or biodiversity for short, as a measure of an ecosystem's value to the planet.

The more species of plants, animals, and other life-forms in a given region, the more resistant that region is to destruction and the better it can perform its environmental roles of cleansing water, enriching the soil, maintaining stable climates, even generating the oxygen we breathe. In 1963 Wilson, along with Robert H. MacArthur, developed the theory called "island biogeography": As the size of a natural area shrinks, the number of species it can sustain shrinks faster. Since then the theory has been confirmed with alarming repetition.

Now 70 and supposedly retired, Wilson is as busy as ever. He continues studying his specialty, ants; he's working on a new classification of one major ant group. He still writes eloquent books; 2 of his 20 titles, *On Human Nature* and *The Ants*, have won Pulitzer Prizes. He still helps guide environmental organizations; currently, he's on the boards of the Nature Conservancy, Conservation International, and the American Museum of Natural History. And he still lectures around the world to save all living species; his Alabama-bred folksiness, combined with a sterling scientific reputation, makes him highly persuasive.

I've followed Wilson's career since 1975, when I wrote about his controversial book *Sociobiology* for *The New York Times.* In those days, he was often attacked for suggesting in that book that certain human behaviors are influenced by urges or instincts that evolved through natural selection. The attacks hurt him deeply. Today sociobiology is widely accepted and, because of his work in biodiversity, Wilson is lionized by many who attacked him decades earlier. "I haven't changed," he observes quietly.

We spoke in his cluttered office at Harvard's Museum of Comparative Zoology, surrounded by stacks of papers, shelves of books, and oversize models of his beloved ants.

**Q:** *As you've helped teach us, we're losing biodiversity at a rate that compares with the great mass extinctions of the prehistoric past. If nothing is done over the next century, how is the earth going to be different?*

**A:** If we continue at the current rate of deforestation and destruction of major ecosystems like rainforests and coral reefs, where most of the biodiversity is concentrated, we will surely lose more than half of all the species of plants and animals on earth by the end of the 21st century.

**Q:** *Is that loss going to happen in isolated patches, or will it be worldwide?*

**A:** Most of the destruction will be from what we today recognize as the "hot spots," where there already is lots of diversity and where habitats are being destroyed. These are primarily but not entirely in the tropics.

**Q:** *A hundred years from now, will America look much different?*

**A:** In the United States the trajectory is less threatening, but even here we would see a shrinkage of fauna and flora over most of the country. And especially in our own hot spots, such as Hawaii and California. For example, in Hawaii alone, where species are disappearing at one of the highest rates in the world, there are more than 100 species of trees that consist of 20 individuals or fewer. So in a century, America would still be biologically rich in most places. But without a stronger conservation policy, it would be partly impoverished, and especially locally a lot of individual states would lose species.

The natural-heritage program of the Nature Conservancy estimates that about 1 percent of native American species of plants and animals have become extinct already, and another 30 percent are in some degree of vulnerability. Even in our national parks, a substantial percentage of mammals have gone extinct—even though they're protected—because the parks are too small.

**Q:** *How will the loss of biodiversity affect human life?*

**A:** On a global basis, I have no doubt at all that there would be severe effects on the quality of life-support systems such as watersheds and air quality and rainfall.

For example, in the Amazon rainforest, a large part of the rain that falls comes from evaporation from the forest itself. As the forest is removed, then a major source of rain is also removed, and substantial parts of the whole Amazon basin could be turned into permanent grassland, with effects radiating out into the breadbasket states of southern Brazil. They would be prone to drought if the Amazon basin dried out.

**Q:** *What has made you so personally interested in this issue?*

**A:** I was a naturalist virtually from childhood, from the age of about nine, when I went out exploring the woods and the fields where we lived. Growing up, especially in Alabama, I would go off on my own, exploring nature because it gave me so much pleasure. I did this right up through my college years—at the University of Alabama and Harvard. The same thing happened later, when I began doing research in the tropics. Especially in the tropics, I became aware of how little of the natural environment has survived.

And so, since my main interest has always been biodiversity, I've been keenly aware that a large part of it was going down the drain. At times, it's made me feel alarmed or depressed.

**Q:** *When did you first become alarmed?*

**A:** In the 1950s, when I was a graduate student and going out to do field research in the tropics. I saw the destruction even then. But I didn't become active in the global environment movement until the late '70s. Up to that time I thought that scientists could pretty well stand aside and let the conservation organizations take care of the activism and building of reserves and restoration of lost habitats. But then I came to realize—especially when others in my field were becoming active in conservation, such as Tom Lovejoy, who is now at the World Bank, and Peter Raven, director of the Missouri Botanical Garden—that scientists had to cross over and get involved because the biodiversity was just disappearing too fast. Conservation organizations did not have the scientific know-how they needed to do planning on a systematic and global scale. That's when I got active.

**Q:** *So did the activist groups change for the better as a result?*

**A:** Yes, but I wasn't solely responsible. I did help change the policy of the World Wildlife Fund–U.S. when I became a member of the board of directors. In the early '80s the World Wildlife Fund changed from an emphasis on specific charismatic species—the giant panda, the leopard, whatever it was—as targets for conservation to biodiversity as a whole. The goal became to save the ecosystem. The effect would be not just to protect the charismatic species but all the rest of the habitat that they need to live. Another shift occurring at that time was also very important. The economic and social welfare of the people who live around the protected areas began to be taken into full account. An environmental program today involves what saving the environment can do for the people, how it can be fitted into the local economy and given value that people immediately understand.

**Q:** *This brings up the question of what individuals can do. Do you think individuals can make choices that matter in the big picture?*

**A:** That's something everybody should do. I would not eat swordfish, for example. It's one of the species driven to commercial rarity. But more important, I think we should be more alert about not buying or using products from species that are protected by CITES [the Convention on International Trade in Endangered Species]. Also, there are a lot of personal habits that, if moderated only in this country, could contribute significantly to saving endangered species.

**Q:** *Like what?*

**A:** Eating imported beef. Before we realized what was happening, the importation of beef from Costa Rica was a significant factor in removing most of its rain-forest. Costa Rica has been essentially stripped of its forest in the past 50 years.

Another good example is coffee-shade-grown versus open-field. Shade-grown coffee is planted among the trees of the natural forest and not in cleared fields. Most aficionados agree that it tastes better. I don't drink coffee, so I'm just quoting. When people ask for shade-grown coffee, they're protecting the forests of Latin America, where a large part of the biodiversity continues to be preserved. If you leave enough of the canopy of the forest and enough of the leaf litter, it's not the same as the original rainforest, but it still has a lot of the biodiversity in it.

**Q:** *Do you think the public has grasped the value of whole ecosystems?*

**A:** Apparently only a minority understand this. The last survey that I know of shows that roughly 20 percent of Americans understand what biodiversity is.

**Q:** *Still, you've said you have hope?*

**A:** I believe that the fate of the world's flora and fauna depends on a combination of science, education, and ethics. We have to get a much better scientific understanding of where biodiversity is and what's happening to it and its value for humanity. And we have to get an understanding of biodiversity into the mainstream of public consciousness so it becomes a principal factor in economic and social policy.

Will this happen? I believe it can, and it must happen soon. The world environment is changing so fast that there is a window of opportunity that will close in as little time as the next two or three decades. I've always thought that we would lose a lot of biodiversity, but how much is hard to say. It could be something like 10 percent of species. But that is far better than the 50 percent or more we will certainly lose if we let things continue as they are today.

# Why Are Biologists Concerned About the Decline of Frog Populations?

Toads and salamanders will probably never catch people's attention as much pandas or whales do. But that does not diminish their importance to the environment.

Why are amphibians—frogs, toads, salamanders, newts, and caecilians—important? For one thing, amphibians play a major role in forest and aquatic ecosystems. Biologist Michael Klemens calls them the "energy cells of this forest ecosystem" because of their role in cycling nutrients and, in turn, providing food for larger animals.[1]

Beyond that, amphibians are considered similar to a miner's canary, the bird carried into mines to check air quality. If the canary died, the miners knew there was a problem with air quality and had to evacuate the mine. Amphibians have been called little environmental sponges, because they easily absorb toxins and harmful rays of sunlight through their porous skin. Amphibians are telling us that there are problems in our environment.

In the United States, there are at least 90 different species of frogs and toads, along with 140 species of salamanders. Today, 27 species of amphibians are listed as endangered or threatened. According the U.S. Geological Survey (USGS), amphibian deformities (extra, misshapen, or missing limbs) have been documented in 44 states and involve nearly 60 species.[2]

Amphibian decline has been called a mystery, especially when extinction takes place in undisturbed areas. Habitat loss, climate change, pollution, disease, and humans hunting or collecting them all play a part. The following section from Raymond D. Semlitsch's book *Amphibian Conservation* highlights some of the key concerns. Raymond D. Semlitsch is a professor of biology and founding director of the Conservation Biology Program at the University of Missouri.

—The Editor

1. United States Geological Survey Factsheet. *Where Have All the Frogs Gone?* 2002. Available online at *http://www.usgs.gov/amphibian_faq.html*.

2. Stutz, Bruce. "Thinking like a Salamander." *OnEarth*. Natural Resources Defense Council, Summer 2004.

## Amphibian Conservation

by Raymond D. Semlitsch, ed.

### WHY ARE WE LOSING AMPHIBIANS?

#### Habitat Destruction and Alteration

Most amphibians depend on both aquatic and terrestrial habitats to complete their life cycle. Conservation of these habitats at local population and landscape levels is critical to maintaining viable populations and regional diversity. However, both habitats have been heavily affected by human use, and these effects are likely the primary cause of reported species declines. Major land-use practices affecting amphibian habitats include agriculture, silviculture, industry, and urban development. These practices are often associated with the filling and draining of wetlands that serve as amphibian breeding sites, removal of trees or natural vegetation in upland habitats used for feeding and as refuges by adults, or alteration of the hydrodynamics of stream and river ecosystems that affect natural functioning and the ability of the ecosystems to support viable populations.

#### Global Climate Change

Global climate changes caused by the accumulation of greenhouse gases and reduction of the ozone layer are now being linked to species declines. This raises serious questions about the future of amphibians in areas of high vulnerability, particularly species with specialized habitat requirements. Two factors of primary concern have been identified: alteration of rainfall and temperature patterns, and increases in ultraviolet (UV-B) radiation. The increase in frequency and severity of El Niño–Southern oscillation events is likely associated with amphibian declines and has been closely related to negative effects on diverse fauna worldwide (e.g., bird communities; reef fish assemblages). Also, increases in temperature and UV-B radiation are likely to interact with other factors, such as disease and chemical contamination.

### Chemical Contamination

Many amphibians encounter chemical contamination in both terrestrial and aquatic environments. Chemicals applied to agricultural fields, golf courses, and forests may directly expose terrestrial juveniles and adults to harmful levels of herbicides, insecticides, and fertilizers. Furthermore, because aquatic environments are the ultimate sink for most chemical contaminants regardless of their source (e.g., agriculture or industry), all aquatic stages of amphibians are likely exposed. In addition, new evidence suggests that airborne contaminants may be affecting amphibian populations in pristine and montane regions that do not receive direct application. Last, research on chemical effects is also exploring sublethal reproductive effects that may have important implications for population persistence.

### Disease and Pathogens

Some amphibian mass mortality events reported in relatively pristine areas of the world have now been linked to infectious diseases. This suggests that pathogens are responsible for declines of some species. Two primary pathogens that appear to be involved are now the current focus of attention: a parasitic chytrid fungus and an iridovirus. What makes both these pathogens so significant to conservation issues is that they can interact with other factors and likely have enhanced susceptibility of various species or populations to other threats.

### Invasive Species

Numerous concerns have been raised about the negative effects of invasive species on native amphibians. Of particular concern are the stocking of predatory game fish, range expansion and introduction of American bullfrogs (*Rana catesbeiana*), and the introduction of exotic species such as cane toads (*Bufo marinus*) and Cuban treefrogs (*Osteopilus septentrionalis*). Fish have been considered the most critical and widespread problem because they can be both competitors and predators of amphibians, especially on aquatic larvae. It is now believed that fish can also act as vectors for disease.

### Commercial Exploitation

The commercial trade in amphibians is a great concern for natural populations and communities for several reasons. First, the direct impact of commercial or illegal collecting may remove a large portion of breeding adults and reduce the capacity of populations to sustain themselves. Second, the reintroduction of wild-collected or captive-reared amphibians (intentionally or accidentally) into natural populations may expose native animals to diseases or pathogens not present in the region (e.g., fish fungus).

## POTENTIAL SOLUTIONS

Interdisciplinary approaches that include social, political, and economic components have been necessary to solve conservation problems in other groups of organisms and must therefore be considered in approaches to help conserve amphibians. Further, looking beyond the local population or ecosystem level to regional or even global scales will be necessary for many of the threats that cross political or geographic boundaries. Many solutions will necessitate international involvement of all stakeholders, including the exchange of information on education, research, and management. Finally, although we can learn much from past efforts to conserve other vertebrate groups (e.g., birds), amphibians have unique physiological, morphological, behavioral, and ecological features that will necessitate different and innovative solutions. . . .

# What Do Declining Bird Populations Tell Us About Our Environment?

You may look for the first robin in the spring, or you may enjoy watching a hawk soar overhead. Whether you enjoy watching birds or not, they do a lot for us. Birds keep down insect populations, pollinate flowers and trees, spread seeds for hundreds of miles, clean up carrion (dead animals), and serve as food for many other animals.

Birds perform one other key service. They tell us things. When they stop singing in the spring, they show that there is a problem with pesticides. Environmentalist Rachel Carson listened, and in 1962, she published a book called *Silent Spring*, which helped launch the modern environmental movement in the United States. When birds sit on their eggs and the eggs crack, this tells us that pesticides are traveling up the food chain, from the insects to the fish that eat them to the birds that eat the fish.

Over the past 200 years, more than 100 species of birds have gone extinct. It is natural for some species to die out. However, several noted biologists have observed that today's extinction rates are occurring at an unnaturally fast rate.[1] The following article from a World Watch Institute 2003 report, "Winged Messengers: The Decline Of Birds," discusses this global conservation issue. Since being founded in 1974 by Lester Brown, the World Watch Institute has provided information on the interactions among key environmental, social, and economic trends. The information is used by government policymakers, businesses, and the general public.

The greatest threat to birds is habitat destruction. Chemical problems, from oil spills to pesticides, as well as climate change, are other threats. In addition, the introduction of exotic species can displace native birds or damage their habitats. Overhunting also threatens birds. There are not enough resources to protect all the habitats, so conservationists have identified key habitats and species on which to focus their efforts.

—The Editor

1. Quammen, David. *Song of the Dodo*. New York: Scribner, p. 607.

## Winged Messengers: The Decline of Birds

by Howard Youth

Fossils reveal that white storks appeared sometime during the Miocene Epoch, between 24 million and 5 million years ago—long before humans, pesticides, power lines, and firearms. The leggy, black-winged birds stalked open, grassy areas and wetlands teeming with insects, frogs, fish, rodents, and other small animals. Over millennia, storks thrived, piling their stick nests on village rooftops. In Europe, villagers wove the graceful, pest-eating birds into lore and legend as baby-carriers and harbingers of good luck. To the south, African villagers called them "grasshopper birds" or "locust birds" because stork flocks snap up the crop-devastating insects while wintering in Africa's Sahel region.

The white stork's fortunes plunged during the 20th century. Food-rich pastures, fallow fields, and wetlands gave way to pesticide-sprayed, intensively managed "modern" farm land that could not sustain the birds. Expanding power line networks added fatal collisions and electrocution to their woes, becoming the greatest direct cause of their mortality in Europe, while in Africa many migrating storks were shot or otherwise caught for food. By the 1980s, white stork populations were declining in all of the Western European countries where they nest.

Over the last decade, though, white stork populations have rebounded, and scientists can't say exactly why. Many biologists suspect the recent wet Sahelian winters. Few believe, however, that storks can have a secure future without careful conservation. The fragile Sahelian plains are becoming degraded from overgrazing and over-hunting, and given the unpredictability of moisture there, along with possible adverse effects of climate change and the other threats mentioned above; the white stork's rebound in the 1990s may prove to be a brief upturn in a long-term decline.

Commuting between continents, white storks respect no political boundaries. They nest, migrate through, or winter in roughly 80 nations. The species' vulnerability, and the international cooperation needed to ensure its survival, exemplify the challenges and promise of future bird (and biodiversity) conservation efforts. The stork's story also highlights how much remains to be learned about the world's feathered creatures, even those that are supposedly well known.

Across the globe, human populations, pollution, temperatures, and introductions of exotic (non-native) species are generally on the rise. Meanwhile, wildlife habitats and water supplies are waning. These trends echo through many bird populations, signaling disturbing global changes. Many of the world's 9,800 bird species are flagging as they struggle against a deadly mixture of often human-caused threats. According to a 2000 study published by a global alliance of conservation organizations called BirdLife International, almost 1,200 species—about 12 percent of the world's remaining bird species—may face extinction within the next century. Although some bird extinctions now seem imminent, many species can still be saved provided we commit to bird conservation as an integral part of a sustainable development strategy. For many reasons, such a commitment would be in humanity's best interests.

Humanity has long drawn inspiration from the beauty, song, and varied behavior of birds. Through the ages, many people believed birds had magical powers and brought good (or bad) luck. Others saw them as guardians, creators, winged oracles, fertility symbols, or guides for spirits and deities. Central America's Mayas and Aztecs worshipped Quetzalcoatl, a dominant spiritual character cloaked in the iridescent green feathers of the resplendent quetzal, a bird now sought by binocular-toting bird-watchers. Ancient Egyptians similarly revered the falcon god Horus and the sacred ibis. Many cultures around the world still ascribe strong spiritual powers to birds, as well as deriving protein and ornaments from them. Native American tribes still incorporate eagle feathers into their rituals, while East African pastoral tribes do the same with ostrich feathers. We also revere birds' flying abilities. Mariners once released ravens and doves aloft in hopes that the birds would steer them toward land, and marveled at the astonishing gliding ability of the albatross.

Inventors, inspired by birds' flight, developed flying machines. Worldwide, artists, authors, and photographers continue to focus their energies on birds, their feathers, and flight.

In habitats around the globe, birds also provide invaluable goods and services. Scientists are just now starting to quantify these behind-the-scenes contributions. Many birds, for example, feed on fruits, scattering seeds as they feed or in their droppings as they flap from place to place. Recent studies revealed that black-casqued, brown-cheeked, and piping hornbills are among tropical Africa's most important seed distributors. In tropical Central and South America, toucans and trogons provide this vital service.

On plains and other open areas, vultures provide natural sanitation services by scavenging animal carcasses. Hummingbirds, orioles, and other nectar-feeding birds pollinate a wide variety of wildflowers, shrubs, and trees, including many valued by people. Meanwhile, thousands of insect-eating species and hundreds of rodent- and insect-eating raptors keep pests in check. In Canadian forests, for instance, populations of wood warblers and evening grosbeaks surge to match outbreaks of spruce budworms, insects that can severely damage forests of spruce and fir. The loss of these birds and their vital ecological contributions tugs at the interconnected fabric of ecosystems.

In addition, many bird species are easily seen or heard, making them excellent environmental indicators. In many cases, they provide scientists with the best glimpse at how humanity's actions affect the world's ecosystems and wildlife. In Europe, biologists consider dippers, round-bodied stream-living songbirds, valuable indicators of clean water because they feed on sensitive bottom-dwelling insects such as caddisfly larvae, which disappear in sullied waters. The disappearance of dippers and their prey also follows water acidification brought on by acid rain or the replacement of native deciduous forests with pine plantations. Other species are important indicators of varied threats to humanity, including chemical contamination, disease, and global warming.

Ornithologists are compiling status reports for all of the world's species, but what they already know is alarming. (See Box 1.) Human-related factors threaten 99 percent of the species in greatest danger. Bird extinctions are on the increase. At least 128 species have vanished over the last 500 years; of these, 103 have become extinct since 1800 and several dozen since 1900. On islands, human-caused bird extinctions are not new: scientists recently concluded that even before European explorers sailed into the region, human colonization of Pacific islands wiped out up to 2,000 bird species that were endemic (found nowhere else). Today, however, people are crowding out bird populations on mainlands as well.

Birds are by no means the only class of animals at risk, of course. Many scientists now consider the world to be in the midst of the sixth great wave of animal extinctions. The fifth wave finished off the dinosaurs 65 million years ago. Unlike previous episodes, however, humans are behind most of the current round of sudden die-offs. One-quarter of the world's mammal species are threatened or nearly threatened with extinction; of the other well-surveyed species, 25 percent of reptiles, 21 percent of amphibians, and 30 percent of fish are threatened.

But if we focus solely on the prospects of extinction, we partly miss the point. From an ecological perspective, extinction is only the last stage in a spiraling degeneration that sends a thriving species slipping toward oblivion. Species stop functioning as critical components of their ecosystems well before they completely disappear. And as conservationists are learning from species reintroduction programs, conserving healthy bird populations now proves far simpler than trying to reconstruct them later.

Although birds are probably the best-studied animal class, a great deal remains to be learned about them, from their life histories to their vulnerability to environmental change. In tropical countries where both avian diversity and habitat loss are greatest—such as Colombia, the Democratic Republic of the Congo (formerly Zaire), and Indonesia—experts just do not know the full scope of bird declines because many areas remain unsurveyed. Species, and distinct populations that may later be considered separate species, may vanish even before scientists can classify them or study their behavior, let alone their ecological importance. Every year, several new bird species

## Box 1. Regional Estimates of Bird Declines

There is no single comprehensive worldwide survey of bird declines, but a broad global picture can be assembled from recent regional surveys, even though they employ varying methodologies:

- A 1994 study revealed that 195 of 514 European bird species (38 percent) had "unfavorable conservation status."
- In 2002, 65 percent of 247 species found in the United Kingdom fell under some category of conservation concern, rating as either "red" or "amber" status. Only 35 percent fell under the "green," or steady and stable, category.
- Based on the North American Breeding Bird Survey's records between 1966 and 1998, some 28 percent of 403 thoroughly monitored species showed statistically significant negative trends. In 2002, the National Audubon Society declared that more than a quarter of U.S. birds were declining or in danger.
- A 2001 BirdLife International study of Asian birds found 664 of the region's bird species (one-quarter of the total) in serious decline or limited to small, vulnerable populations.
- Some Australian ornithologists believe that half of their island nation's land bird species, including many endemic parrots, could become extinct by the end of the century, although recent breeding bird surveys chronicled little difference in status for most species over the past 20 years.

are described. One of the century's first was an owl discovered in Sri Lanka in January 2001, the first new bird species found there in 132 years. Other species, while known to science, have not been seen in years but may still survive. These scarce birds sit at a crossroads, as does humanity. One path leads toward continue biodiversity and sustainability. The other leads toward extinction and imbalance.

Section B

# Endangered Species

# Is the Concern Over Whale Populations a New Issue?

Whales travel in all oceans, as can the whaling ships that hunt them. Over the last century, many species of whales were hunted to near extinction. Today, about half (seven) of the great whale species are listed as endangered or threatened. The problem is not new. In fact, countries became concerned over 50 years ago about the exploitation of whales. In 1946, nations joined to form the International Whaling Commission (IWC) "to provide for the proper conservation of whale stocks and thus make possible the orderly development of the whaling industry."[1] The following World Wildlife Fund (WWF) article outlines the history of the IWC. The World Wildlife Fund works around the world to create marine sanctuaries and to demonstrate the economic value of whale-watching. WWF is campaigning to have at least at least 10% of marine areas under some form of protection by 2012.

In 1986, the IWC declared a moratorium (suspension of activity) on whaling, in an effort to give the whale populations a chance to recover and to complete "an in-depth evaluation of the status of all whale stocks in the light of management objectives and procedures . . . that . . . would include the examination of current stock size, recent population trends, carrying capacity and productivity."[2] It is still in effect today.

As is the case with other large mammals, whales are slow to reproduce. Today, several populations of whales remain "highly endangered" with a total number of 500 or less. The North Atlantic right whale, the Western North Pacific gray whale, and the blue whale are examples.[3] Other whale populations, including the humpback whale, are rebounding more quickly.[4] Despite the known dangers to the whale populations, three countries—Japan, Norway, and Iceland—continue to carry out whaling operations using loopholes found in the Whaling Convention. Since 1986, more than 25,000 whales have been killed by commercial whalers.

As of July 2004, there were 57 nations participating in the IWC, with about an even split between whaling and non-whaling nations. The IWC is the only international regulatory group that manages whale populations. In 2003, the IWC established a Conservation Committee, an important shift in focus toward whale conservation and research efforts. These programs take on great significance, since the threats to whales now come not only from whalers. Pollution, by-catching (catching unwanted animals in fishing nets), and the effects of global warming are cause for additional concerns.

—The Editor

1. International Whaling Commission Factsheet. *IWC Information*. 2004. Available online at *http://www.iwcoffice.org/commission/iwcmain.htm.*
2. International Whaling Commission Factsheet. *Whale Population Estimates*. 2004. Available online at *http://www.iwcoffice.org/conservation/estimate.htm*.
3. International Whaling Commission. *2004 Press Release from the International Whaling Commission's 56th Annual Meeting.* Available online at *http://www.iwcoffice.org/other/newadditions.htm*.
4. International Whaling Commission Factsheet. *Whale Population Estimates*. Available online at *http://www.icrwhale.org/eng/estimates.pdf*.

## Whaling

from the World Wildlife Fund

### WHERE DID THE IDEA OF THE INTERNATIONAL WHALING COMMISSION COME FROM AND WHY?

Whaling as an industry began around the 11$^{th}$ Century when the Basques started hunting and trading the products from the northern right whale (now one of the most endangered of the great whales). They were followed first by the Dutch and the British, and later by the Americans, Norwegians and many other nations. Humpback and sperm whales were the next targets of commercial whaling, with oil for lighting and other uses as the most important product. In the late nineteenth century the whaling industry was transformed by the development of steam powered ships, enabling the hunting of faster blue and fin whales, and of the explosive harpoon, enabling further reach and increased accuracy. The new technology, coupled with the depletion of whales in the rest of the world, led to the spread of hunting to the Antarctic, where huge concentrations of feeding whales made large-scale whaling highly profitable. The First World War provided a large market for explosives using glycerine from baleen whale oil provided by British and Norwegian whaling in the Antarctic. Meanwhile Japanese whaling had developed separately as a coastal industry, mainly for humpback, right and grey whales.

Since whales migrate world-wide through both coastal waters and the open oceans, the need for international cooperation in their conservation became evident. By 1925, the League of Nations recognised that whales were over-exploited and that there was a need to regulate whaling activities. In 1930, the Bureau of International Whaling Statistics was set up in order to keep track of catches. This was followed by the first international regulatory agreement, the Convention for the Regulation of Whaling, which was signed by 22 nations in 1931. However, some of the major whaling nations, including Germany and Japan, did not join and 43,000 whales were killed that same year.

With species after species of the great whales being hunted close to extinction, various nations met throughout the 1930s attempting to bring order to the industry. Finally, in 1948 the International Convention for the Regulation of Whaling (ICRW) came into force. The Preamble states that "Recognising the interest of the nations of the world in safeguarding for future generations the great natural resources represented by the whale stocks. . . . having decided to conclude a convention to provide for the proper conservation of whale stocks and thus make possible the orderly development of the whaling industry." The International Whaling Commission (IWC) was established as its decision-making body, originally with 14 member states. The IWC meets annually and adopts regulations on catch limits, whaling methods and protected areas, on the basis of a three-quarters majority vote. In recent years the IWC, recognising new threats to whales, has moved towards a broader conservation agenda which includes incidental catches in fishing gear and concerns related to global environmental change. Whale hunting by indigenous people, called "aboriginal subsistence" whaling, is subject to different IWC controls than those on commercial whaling.

Today the IWC has 52 member states, including whaling countries, ex-whaling countries, and countries that have never had whaling industries but joined either to have a voice in the conservation of whales or to support whaling interests.

## SUCCESSES AND FAILURES OF THE IWC

For the first 15 years of its existence the IWC acted as a "whalers club" and imposed hardly any effective restrictions on whaling. Catch limits were set far too high and, since the IWC lacks a compliance and enforcement programme, were often exceeded. These management shortfalls resulted in the continued depletion of species after species. In particular, huge declines occurred in the Antarctic, where in the 1961/62 season, the peak was reached with over 66,000 whales killed. By then, however, it was becoming increasingly hard for the whalers to find enough whales to kill. From a pre-whaling population of about 250,000 blue whales in the Southern Hemisphere, there are now estimated to be fewer than 1,500 remaining.

Also in 1961, WWF [the World Wildlife Fund] was founded and accepted the challenge of reversing the declines in whale populations. "Save the whales" campaigns spread around the world, promoting calls for whale sanctuaries and a moratorium on commercial whaling (most notably by the UN Conference on the Human Environment in 1972). Instead of implementing a moratorium, in 1974 the IWC adopted a New Management Procedure (NMP), designed to set quotas on the grounds of scientific assessments and sustainability. However, the NMP was not precautionary at all; it depended on having much more information on whale stocks than was available, quotas were still set too high, compliance was still lacking, and whale populations continued to decline.

At the 1979 IWC meeting, a moratorium on all whaling using factory ships (with an exception for minke whales) was agreed. The IWC also declared the entire Indian Ocean as a whale sanctuary. From then on, successful non-lethal whale research took place in that area (some of it funded by WWF). However, it was also revealed that the USSR [Soviet Union] had been falsifying reported numbers and species caught on a massive scale, with the meat being sold to Japan. Conservation concerns expressed by scientists, WWF and other conservation organizations and conservation-minded governments grew deeper.

At the 1982 IWC meeting, a proposal for a moratorium on all commercial whaling, to come into force in 1986, was tabled by the Seychelles. The vote was comfortably won with a majority of 25 to 7, with five abstentions. Japan, Norway, and the USSR subsequently lodged official objections giving them exemption from the moratorium, but Japan withdrew its reservation as of the 1987/88 season.

Because of the problems with the New Management Procedure, the IWC asked its Scientific Committee to produce a fail-safe management system that could ensure that any future commercial whaling would never again deplete whale stocks. In 1994, the Revised Management Procedure (RMP), a set of precautionary rules for setting catch limits, was agreed by IWC Resolution, although not formally adopted into the IWC "Schedule," or rules of operation. The RMP is designed as one part of a Revised Management Scheme (RMS) which would also include rules for conducting surveys of whale numbers and for the inspection and observation of commercial whaling. Continued controversy regarding the need for additional safeguards that would prevent any repetition of past abuses has so far prevented the adoption of the RMS.

In 1994, after an intensive campaign by WWF and other NGOs [nongovernmental organizations], the 50 million square kilometer Southern Ocean Whale Sanctuary came into force. In the long term this should ensure the recovery of the world's whale populations that have suffered most from exploitation. However, although several countries initiated non-lethal research in the Southern Ocean Sanctuary, Japan is still conducting lethal so-called "scientific" whaling within the boundaries of the Sanctuary, as well as in the North Pacific.

## THE CURRENT SITUATION IN THE IWC

Over recent decades, the IWC has taken some encouraging steps in changing its emphasis towards conserving and studying whales, most recently in 2003 with the establishment of a Conservation Committee. However, the whaling nations of Japan, Norway and Iceland retain politically influential whaling industries that wish to carry on whaling on as large a scale as possible. All three countries are exploiting loopholes in the Whaling Convention in order to kill more than 1,200 whales each year in spite of the IWC's moratorium on whaling. Norway hunts whales under its objection to the moratorium, and Japan has been whaling under the guise of "scientific research." . . . Most recently, Iceland joined the IWC with a formal objection to the moratorium and, although claiming they would not undertake commercial whaling before 2006, immediately began a "scientific whaling" program. The current membership of the IWC is approximately evenly divided between whaling and non-whaling nations, resulting in a political deadlock which makes it impossible to secure the 3/4 majority of votes needed to make major changes. All in all, whaling is taking place and increasing yearly without any international control.

While the debate has raged over how best to manage commercial whaling, emerging threats to the future of all cetacean populations have begun to be addressed by the IWC, both within its Commission and its Scientific Committee. Among the important conservation issues under consideration have been: conservation of "small" cetaceans; incidental catches in fishing gear (by-catch); whale watching; protection of highly endangered species and populations; whales and their environment (including toxic chemicals and other marine pollution); ecosystem management concerns; sanctuaries; enforcement and compliance; management of "scientific whaling;" and collaboration with other organizations. These issues, of critical importance to the future of all cetaceans, now constitute a broad and growing, although controversial, conservation agenda within the IWC.

### THE IWC IN THE 21st CENTURY

The IWC's mandate requires first and foremost that it prevent the return of uncontrolled large-scale commercial whaling. The 1946 ICRW, however, was negotiated at a time before the broad range of threats to cetaceans were understood or even recognised to exist—tied not only to an era which had little understanding of the complex web of marine issues facing all cetaceans, but also to a very different political era than the one in which it exists today. In the more than 50 years since the Convention text was adopted, it has become impossible to separate the threats presented by commercial whaling from

## Key Dates

| | |
|---|---|
| **1925** | League of Nations recognises overexploitation of whales |
| **1930** | Bureau of International Whaling Statistics set up |
| **1931** | First international regulatory agreement |
| **1931** | Bowhead whale protected |
| **1935** | Northern & Southern Right whale protected |
| **1937** | Grey whale protected |
| **1946** | International Convention for the Regulation of Whaling agreed |
| **1949** | International Whaling Commission established |
| **1961** | Highest known number of whales killed (66,000) |
| **1963** | Humpback whales protected in Southern Hemisphere |
| **1967** | Blue whales protected in Southern Hemisphere |
| **1979** | Indian Ocean Sanctuary established moratorium on factory ship whaling (except for minkes); Sei whales protected (some exceptions) |
| **1981** | Sperm whales protected (some exceptions) |
| **1982** | IWC moratorium on commercial whaling agreed |
| **1986** | IWC moratorium on commercial whaling comes into force |
| **1987/8** | Japan begins scientific whaling |

those of marine pollution, commercial by-catch, or over-fishing. It is far preferable, and of greater potential conservation to cetaceans, to now address all of the threats to cetacean populations in a broad, multilateral context, as the IWC has begun to do. The ICRW is currently the only international instrument available to formally address all cetaceans and all threats to their continued existence. WWF believes the IWC must continue to expand its scope to address the other human activities which threaten cetaceans and focus action on ensuring the survival of the most threatened species.

**1989** Lowest known number of whales hunted (326)

**1992** North Atlantic Marine Mammals Commission (NAMMCO) established

**1993** Norway resumes commercial whaling under objection to the moratorium

**1994** Southern Ocean Whale Sanctuary established

**1994** RMP approved

**1997** Numbers of minke whales killed for commercial and "scientific" purposes by Japan and Norway increase to over 1,000 animals

**2000** Japan extends its scientific whaling to include Bryde's and sperm whales as well as Northern and Southern minkes

**2001** Norway announces the resumption of international trade in whale meat and blubber (although this does not take place)

**2002** Japan extends its scientific whaling to include endangered sei whales

**2002** (October) Iceland succeeds in rejoining IWC with a reservation to the moratorium, valid after 2006

# What Are Countries Around the World Doing to Protect Endangered Species?

One important international tool in use today to protect endangered plants and animals is CITES (Convention on International Trade in Endangered Species). The goal of CITES is to stop any international trade of wild animals and plants that threatens their survival. It came out of growing awareness in the 1960s that some organized action was needed to protect certain plants and animals from extinction. In 1963, a resolution at a meeting of the World Conservation Union was developed and finally adopted by 80 countries in 1973. Then, in 1975, CITES went into effect. Today, it has 166 voluntary member countries, making CITES one of the largest conservation agreements in the world.[1]

About 5,000 species of animals and 28,000 species of plants are protected by CITES. Member countries work to stop the billion-dollar trade in illegal plants and animals by following CITES recommendations. As an example, in August 2004, authorities in Indonesia sent a strong message by convicting five poachers to prison terms and large fines for illegally trading endangered Sumatran tigers, one of the animals protected by CITES.

—The Editor

1. CITES Factsheet. *What is CITES?* 2004. Available online at *http://www.cites.org/eng/disc/what.shtml*.

## CITES

from the World Wildlife Fund

For three decades, the Convention on International Trade in Endangered Species of Wild Fauna and Flora, known as CITES, has been the largest and by some accounts, the most effective international wildlife conservation agreement in the world.

The treaty entered into force in Washington DC in 1975, in response to concerns that many species were becoming endangered because of international trade. Because this trade crosses national borders, international collaboration and co-operation is crucial to ensure this trade is sustainable and controlled and does not threaten or endanger wildlife.

Since the Convention entered into force, more than 30,000 species of animals and plants have been listed on its Appendices, from tigers and elephants to mahogany and orchids.

### HOW DOES CITES WORK?

CITES regulates international trade in species by including species on one of three Appendices.

- **Appendix I** bans commercial trade in species threatened with extinction.
- **Appendix II** regulates international trade in species whose survival in the wild may be threatened if levels of trade are not regulated.
- **Appendix III** is a list of species included at the request of a Party (a country that has joined CITES) that needs the cooperation of other countries to help prevent illegal exploitation.

### WHAT IS THE CoP?

Every two to three years, Parties to CITES meet to discuss a variety of issues. This meeting is called a meeting of the Conference of the Parties—CoP. CoP 14 is likely to take place in 2007.

### WHAT ARE THEY FOR?

At each CoP, CITES Parties discuss proposals to amend the Appendices. Parties vote on each proposal. Two thirds of the Parties that are present and voting must vote in favour of a proposal for it to be accepted.

Parties also discuss and come to agreement on a range of Resolutions and Decisions that may relate to, for instance, the interpretation of the Convention, its operation for particular taxa or specimens, or specific trade-related conservation measures.

Only Parties can vote. NGOs [nongovernmental organizations] such as WWF [World Wildlife Fund] and TRAFFIC provide technical and scientific advice, and are able to participate and speak at the CoP. WWF has actively participated in all previous CITES CoPs, and brings to the table both its policy expertise, and its 40 years of field expertise in more than 60 countries.

## CITES: WHAT HAS BEEN ACHIEVED SO FAR?

### A Viable and Working Convention for Conservation

Since CITES came into force, the convention has banned international trade in rhino horn and helped to ensure that rhinos continue to survive in the wild.

CITES also banned international trade in ivory in 1989 to combat a massive illegal trade in ivory which caused dramatic declines in elephant populations throughout most of Africa in the 1970s and 1980s.

The ban was successful in eliminating some of the major ivory markets, leading to reduced poaching and allowing some populations to recover.

Other measures adopted by CITES have led to improvements in the management and regulation of trade in a myriad of other species such as sturgeon caviar, some species of sharks, seahorses, crocodiles.

## WHAT SPECIES ARE ALREADY ON CITES LISTS?

Around 5,000 species of animals and 25,000 species of plants are protected by CITES. To see the full list visit the CITES Website at *http://www.cites.org.*

### CITES Successes at CoP 12

WWF helped secure protection for a range of species including

- whale and basking sharks
- big-leaf mahogany
- seahorses. . . .

# How Are Endangered Species Protected in the United States?

Another tool for protecting endangered animals was the Endangered Species Act (ESA) of 1973, passed by the U.S. Congress to replace a similar act passed in 1969. As President Richard Nixon said upon signing the law, "Nothing is more priceless and more worthy of preservation than the rich array of animal life with which our country has been blessed."[1]

The program is administered by the U.S. Fish and Wildlife Service (FWS) and the National Oceanic and Atmospheric Administration (NOAA) Fisheries. After a species is listed as endangered or threatened, a recovery plan is prepared. The recovery of the bald eagle, the American national symbol, is one the ESA's greatest success stories. Today, the FWS considers more than 500 of the listed species to be stable or improving in status. However, there are still more than 1,200 species in danger—that leaves about 700 species, over half of the total, in a condition that is neither stable nor improving.

Does ESA work? Of all the species listed between 1968 and 2000, only seven, which is less than 1%, have been declared extinct.[2] In addition to the bald eagle, success stories include the peregrine falcon and the gray wolf.

But do we need it? In 1995, the National Research Council issued a report evaluating the ESA. It concluded that "there is not doubt that it has prevented the extinction of some species and slowed the decline of others."[3]

The following is a section from a 2003 U.S. Fish and Wildlife Service Report entitled *Endangered Species Bulletin* No. 4., Vol. XXVII. The complete report is available online at the U.S. Fish and Wildlife Service's Website.

—The Editor

1. National Oceanic and Atmospheric Administration (NOAA). *Fisheries: Office of Protected Resources: Endangered Species Act of 1973* Factsheet. Available online at *http://www.nmfs.noaa.gov/prot_res/laws/ ESA/ESA_Home.html.*

2. U.S. Fish and Wildlife. *Endangered Species Bulletin No.* 4. July–December 2003.

3. Union of Concerned Scientists. *Backgrounder: Biodiversity Baloney: Some Popular Myths Undone*. 2002. Available online at *http://www.ucsusa.org/global_environment/archive/page.cfm?pageID=393*.

## Endangered Species Bulletin No. 4.

from the U.S. Fish and Wildlife Service Report

The purpose of the Endangered Species Act is to conserve endangered and threatened species and the ecosystems upon which they depend. The ultimate symbolic action in a species' recovery effort is taking the species off the endangered and threatened species list because it is no longer threatened with extinction now or likely to become so within the foreseeable future. Final delisting and downlisting (i.e., changing a species' status from endangered to threatened) is achieved through time, steadfast dedication, and the use of existing and innovative techniques.

In the Midwest, for example, the prairie bush clover *(Lespedeza leptostachya)* has been helped by years of dedication toward recovery. Restoring the prairie bush clover focused on identifying and protecting populations in both the core and peripheral portions of its range. All that remains before delisting is to conduct a viability analysis of the protected populations to ensure that they will remain healthy.

The endangered Magazine Mountain shagreen (*Mesodon magazinensis*) is restricted to a single population found on the talus slopes of Magazine Mountain in the Ozark National Forest of Arkansas. Evidence has revealed that the range of this snail had not contracted; instead, it has always been endemic to this one site. As part of the construction of a state park on Magazine Mountain in 1995, the U.S. Forest Service began monitoring the snail for 10 years. At the end of this period, if the shagreen is still stable, the species could be considered for delisting. The final survey will be conducted in the spring of

2005, but at this time the results of the survey indicate that the population has remained stable.

One of the most recognized species on the list of endangered and threatened species is the gray wolf (*Canis lupus*). After decades of widespread persecution of the wolves due to perceived and real conflicts between wolves and human activities, it is estimated that only several hundred wolves survived in northeastern Minnesota and on Isle Royale, Michigan, with possibly a few scattered wolves in the Upper Peninsula of Michigan, Montana, and the American Southwest at the time the Endangered Species Act of 1973 was enacted.

Today, with improved and coordinated management, the introduction of wolves back into areas where they once existed, and the cooperation of the states, conservation organizations, many private landowners, and numerous other partners, gray wolf populations have re-bounded in the East to over 3,000 wolves. In the Northern Rocky Mountains, there are an estimated 664 wolves in 44 packs in northwestern Montana, Idaho, and in and around Yellowstone National Park. Populations in both regions are exceeding their numerical recovery goals. As a result, in April 2003, the Service downlisted the gray wolf from endangered to threatened in the Eastern and Western Distinct Population Segments (the Southwest DPS is still listed as endangered) and established two new special rules under section 4(d) of the ESA that increases our ability to respond to wolf-human conflicts in these areas. At the same time, the Service announced its intention to propose delisting the gray wolf in the Eastern and Western DPSs within the near future. Another strong sign of its recovery progress.

Many of the other articles in this issue of the *Bulletin* describe the dedication and resolve required to achieve recovery of a species, including an article on the riparian brush rabbit (*Sylvilagus bachmani riparius*), which would have gone extinct if the Service hadn't taken action, and Robert "Sea Otter" Jones' efforts to recover the Aleutian Canada goose (*Branta canadensis leucopareia*).

Recovery is on the horizon for many species on the list. Ninety-seven percent of U.S. species listed as of September 30, 2002, still survive and many of them are headed toward recovery. In fact, the Service considers over 500 listed species to be stable or improving in status. By any measure, this is a tremendous success. The many partners involved in contributing to recovery deserve the credit. Endangered or threatened species recovery is often a long, slow process, but the goal of preventing extinction and giving hope to other listed species is attainable. . . .

## SAVING SPECIES ON THE BRINK OF EXTINCTION

According to paleontologist Niles Eldredge, Earth is experiencing its sixth major wave of extinction. Our nation has not escaped the forces threatening plant and animal species. Of the more than 1,200 species in the United States currently protected by the Endangered Species Act, 417 are declining in number and 28 others are now believed to be extinct. Many critically endangered species are geographically concentrated in "hot spots."

Aquatic species in the Southern Appalachian and Lower Tennessee Cumberland ecosystems. The south-eastern U.S. has the greatest diversity of freshwater mussels and crayfishes in the world, and the highest diversity of freshwater fishes and snails in the country. Conservatively, we estimate that nearly 40 of these species have reached such low population numbers that a single isolated event could cause their extinction. Because many of these species survive in only a fragment of their former range, a single catastrophic event could cause their extinction. Among the southeastern aquatic species that are critically endangered are the tan riffleshell (*Epioblasma florentina walkeri*), with only one reproducing population in Indian Creek, Virginia; the plicate rocksnail (*Leptoxis plicata*), currently found only in the Locust Fork, Alabama; and the boulder darter (*Etheostoma wapiti*), found only in the Elk River in Tennessee and Alabama. Recovery actions needed to save these species include developing propagation technology, restoring habitat, reintroducing the species into restored habitat, and supporting sustainable development and resource use that also conserves the species.

## ENDEMIC HAWAIIAN PLANTS AND ANIMALS

Hawaii has more critically endangered species than any other state. As of October 24, 2003, there are 312 listed species, 106 candidate species, and over 1,000 species of concern. Of these, there are 102 endangered species, including 11 birds, four tree snails, and 87 plants, in such low numbers that could be rendered extinct by a single isolated incident, such as a fire or hurricane. The most serious threats to these species include the continued influx of competitive and predatory nonnative species, and the fragmentation and degradation of habitats. Efforts needed to save these species include removing or controlling destructive invasive species. . . . Emergency management needed to protect Hawaii's critically endangered species will also benefit other listed species and at least 30 candidate species. While the Southern Appalachians and Hawaii host groupings of critically endangered species, there are many other such species across the Nation. Some examples follow:

The Carson wandering skipper (*Pseudocopaeodes eunus obscurus*) is a butterfly currently known from only two populations, one in Washoe County, Nevada, and one in Lassen County, California. It needs grassland habitats on alkaline substrates to survive, and this habitat type has been reduced by activities associated with development, certain agricultural practices, collection, and nonnative plant invasions. This rare butterfly is also threatened by unscrupulous collectors.

The pallid sturgeon (*Scaphirhynchus albus*) is a fish that has survived for over 200 million years but it is now on the verge of extinction. After the construction of dams on the Missouri River, the ecosystem inhabited by pallid sturgeon was almost completely altered. There is limited evidence that reproduction is still occurring in the wild. Most of the pallid sturgeon in the wild are 40 to 50 years old. The window of opportunity for obtaining reproduction from these individuals is close to the end. Retrofitting Missouri River fish hatcheries to accommodate the needs of this unusual species is critical to augmenting the wild populations. The efforts the Service and our partners make during the next five years will be crucial for preventing this species' extinction.

The Mississippi gopher frog (*Rana capito sevosa*) was once found in suitable habitat within the Lower Coastal Plain from Florida to eastern Louisiana. Today, however, the frog is known from only one small pond in extreme south-central Mississippi. It spends most of the year underground, often using the burrows of the threatened gopher tortoise (*Gopherus polyphemus*). In spring, the frogs travel overland to reach small ephemeral ponds, where they mate and lay eggs. Most of these ephemeral ponds have been lost to forestry practices, agriculture, and, in some cases, conversion to permanent ponds stocked with game fish. Surrounding habitats with gopher tortoise burrows have likewise been lost to development and land use changes. Preventing the extinction of this unique frog will require the restoration of ponds and surrounding habitats and the reintroduction of frogs from the surviving population.

The emergency-listed Columbia Basin pygmy rabbit (*Brachylagus idahoensis*) has fewer than 50 individuals in the wild, all in Douglas County, Washington. It faces imminent extinction resulting from disturbances to its sagebrush habitat, disease, predation, and loss of genetic diversity. We need to develop a program to breed the rabbits in captivity for release into the wild. Its survival will depend on working with our partners and stakeholders to implement conservation actions and to integrate these actions with agricultural practices.

Attwater's greater prairie-chicken (*Tympanuchus cupido attwateri*) may be North America's most endangered bird. Since 1996, captive-bred birds have been released on the Attwater's Prairie-Chicken National Wildlife Refuge and the Texas City Prairie Preserve. However, these sites can support only a small number of prairie-chickens. Saving this species will require strong partnerships with private landowners. Prescribed burns, brush control, conversion of land back to native grasses and forbs, and grazing regimes that will foster native species are needed.

Halting the loss of these and other species will require continued collaboration between the Service and our many partners. By working together, we can conserve the remaining habitats and restore others, while at the same time supporting sustainable development and land use.

# What Dangers Do Wolves Still Face in Our Country?

The wolf has long been celebrated by Native Americans for its hunting prowess and cunning. But as the early American settlers traveled west, wolves were seen as a threat to their livestock and way of life. As an answer, the government made wolves the target of a hugely successful campaign to destroy the species. By the time the Endangered Species Act (ESA) was passed in 1973, there were no wolves left in the continental United States, except for a small population in northern Minnesota. Even today, the wolf is back in only about 2% of the land it once roamed.

The following article discusses the issues surrounding the recovery of wolves in the United States. Polls indicate that many Americans support the return of the wolf. And research continues to show that the presence of top predators, such as wolves, help keep an ecosystem healthy.

The ESA states that threatened or endangered "fish, wildlife, and plants are of esthetic, ecological, educational, historical, recreational, and scientific value to the Nation and its people."[1] The wolf is one of the ESA's most successful recovery cases (even though it is only back to 2% of its original range). There are now three areas in which wolves are found in the continental United States: parts of the east, west, and southwest.

As the following article explains, conservationists are concerned about what happens when a species status is changed, or "downlisted" from endangered to threatened, or taken off the list completely. If they were removed from the lists, wolves would lose federal protection, and would fall prey to state laws, some of which still echo the wolf-killing philosophy of the past. In 2004, two laws were introduced in Wyoming that would greatly curtail wolf protection once wolves are de-listed. Fortunately for the wolves, the bills were defeated. But Alaska, a state in which wolves are not an endangered species, still uses controversial air-borne hunting to kill wolves.[2]

The challenge is for the U.S. Fish and Wildlife Service to continue the successful reintroduction and protection program it has begun. The wolf fits the aesthetic, ecological, educational, historical, recreational, and scientific values identified in the ESA. The job of saving the wolves is far from being done.

—The Editor

1. Endangered Species Act of 1973. Section 2 (a) (3). Available online at *http://endangered.fws.gov/esaall.pdf*.
2. Defenders of Wildlife. *Wolves in Alaska: Overview*. July 2004. Available online at *http://www.defenders.org/wildlife/wolf/Alaska.html*.

## America's Wolves Threatened Again

by Nina Fascione

The future of wolves is once again at a crossroads in the lower 48 states. The U.S. Fish and Wildlife Service (FWS) released a rule in March [2003] that significantly reduces federal protections for wolves and sets the stage for removing them from Endangered Species Act (ESA) protections entirely. Conservationists are particularly concerned with two aspects of the new rule: the absence of any plans to pursue wolf recovery in additional areas and the fear that recovered wolf populations will suffer under state management.

Until now, FWS has done an exceptional job at restoring wolves to the United States. The reintroduction of wolves in Yellowstone National Park and central Idaho is considered by many to be one of the greatest conservation successes of the 20th century. That wolf population has grown from the 66 reintroduced by FWS biologists in 1995 and 1996 to more than 600 today. And under federal protection, wolves in the Great Lakes region have increased from fewer than 1,000 in the 1970s to more than 3,000 today. Still, with approximately 4,000 gray wolves in the Lower 48 states, the species has been returned to less than two percent of its historic range. The new FWS rule lets the agency stop short of completing the job—essentially halting all efforts to continue wolf recovery.

The rule addresses three recovery areas in the United States—the West, the East and the Southwest. In the West, the reintroduced wolves will retain their current protection under the ESA; beyond that population, wolves will be downlisted from endangered to the less protected status of threatened throughout the Rockies and the Pacific Northwest. Although only three of nine states in the region with vast areas of suitable habitat have seen recovery efforts, FWS says it has reached its recovery goals in the northern Rockies and therefore the job is done in the West. To the dismay of conservationists, further recovery in California, Utah, Oregon and Washington is not being considered.

Similarly, wolves will be downlisted to threatened in the Great Lakes and Northeastern states. Despite the absence of wolves outside of Minnesota, Michigan and Wisconsin, no new recovery areas will be pursued and FWS plans to eventually remove the Great Lakes wolf population from federal protection. The final recovery region is the Southwest, for which FWS has yet to develop a recovery plan with specific goals for the reintroduced Mexican gray wolf. These wolves will retain their endangered status and are classified as an experimental population.

Another troubling aspect of the proposed rule is that the withdrawal of federal protections may place wolf populations that have already been restored in jeopardy again. Without federal oversight, management of the species would be turned over to state agencies. In preparation, the states have been busily preparing wolf plans, a required blueprint for how states with existing wolf populations will manage the species. Many state agencies have already demonstrated that they are not willing or capable of taking the steps necessary to further wolf restoration.

Unfortunately, the myths and superstitions that led to persecution of wolves in past centuries still survive today in the minds of many. It is not uncommon for those that promote anti-wolf rhetoric to maintain influential positions in local, regional and state legislatures. In Idaho, state lawmakers passed a resolution in 2002 stating that the legislature "not only calls for, but demands, that wolf recovery efforts in Idaho be discontinued immediately, and wolves be removed by whatever means necessary." The approved Idaho wolf management plan reiterates that the resolution, though it does not carry the weight of the law, continues to represent the official position of the state.

Currently there are 21 pieces of state legislation pending in the West that, in the absence of federal oversight, could all negatively impact current and future wolf recovery. Colorado still has a wolf bounty on the books. Two years ago, Minnesota developed a state management plan that includes, in addition to liberal control actions in parts of the state, a $150 payment to "animal controllers" for killed wolves, eerily resembling the bounties of old that contributed to the species' decline in the first place. Although the bill did not have popular support, the Minnesota legislature was able, after multiple efforts over several years, to attach the wolf management plan to a more popular piece of natural resource legislation and pass it as part of a larger package. Environmentalists sued to have this "log-rolled" legislation thrown out, but they lost in court.

These measures do not give conservationists much confidence that wolves will be protected or further restored once federal protections are removed. According to William Snape, chief counsel at Defenders of Wildlife, "Prematurely delisting wolves in the lower 48 makes no sense biologically, and it violates the recovery mandate of the Endangered Species Act. Just because we are better off today than we were a decade ago doesn't mean the job is over. Handing management of wolves back to the states, who have so far shown no intent to responsibly conserve predators, would literally turn the clock back 100 years to when government officials actively promoted the wolf's demise."

It's not only wolves that are at risk from FWS's new rule; entire ecosystems may lose the benefits brought on since wolf restoration began in the mid 1990s. While tourists flock to Yellowstone National Park's Lamar Valley to witness for themselves the park's most glamorous predator, forest researchers are equally fascinated by the gradual return of quaking aspen, one of the most ecologically important riverside tree species in the Yellowstone ecosystem. Ever since gray wolves were reintroduced in the park seven years ago, aspen have experienced a dramatic comeback. Scientists have determined that there is a direct link between the recovery of the ecosystem's top predator and the trees. Aspen growth essentially stopped once wolves were removed from the park in the 1920s, says William Ripple of Oregon State University's department of forest resources. Now that the wolves have returned, the trees are growing again. Elk, as it turns out, forage differently depending on whether predators are present. During the 60 years wolves were absent from the park, elk spent more time browsing alongside rivers, trampling the low vegetation and inhibiting new growth of native tree species, including aspen. With wolves once again on the scene, elk are behaving more cautiously—avoiding areas with dense foliage and spending more time in open areas in order to keep an eye on their surroundings. This behavior change is altering the entire landscape.

The changes will continue to spread. In a system known as a "trophic cascade"—the interactions between different levels of the food chain—predators exert an influence on more than just the numbers of their direct prey. By altering the movements and foraging behavior of elk, wolves are playing a key role in preserving the integrity of Yellowstone's overall biodiversity. "If the aspen and other riparian vegetation of Yellowstone continue to grow taller and expand in canopy cover, the numerous benefits to ecosystem processes will include stream channel stabilization, flood plain restoration and higher water tables. Through a trophic cascade effect of improved habitat, wolves may be beneficial to numerous species of vertebrates and invertebrates such as fish, birds, beaver and butterflies, as well as many other species of wildlife," Ripple says.

Ripple's research on aspen supports what biologists and conservationists have been saying for years—that top predators, particularly wolves, provide an essential service to the environment. "Wolf reintroduction may be useful for programs designed to restore riparian areas and biodiversity, and should be considered for other areas of the United States, as well as other areas of the world where wolves once roamed," Ripple says.

Other beneficial changes have been documented since wolf reintroduction into Yellowstone. Wolves have reduced coyote populations in some areas by up to 50 percent, which in turn enabled smaller animals such as foxes and rodents to rebound. With more rodents available, birds of prey have thrived. Leftover wolf kills benefit a host of other species, including ravens, magpies, golden and bald eagles, foxes, cougars, insects and even the park's famous grizzly bears.

Because of the important role wolves play in the ecosystem, biologists and environmentalists have long encouraged FWS to pursue wolf restoration possibilities in many regions. Several areas of the country have been identified as possessing suitable wolf habitat but still lack this top carnivore. Based on modeling and other scientific studies, biologists feel confident that portions of the Pacific Northwest, the Northeast and the southern Rockies could, combined, be home to 4,000 wolves or more. This is double the number of wolves currently in the lower 48 states.

The Klamath-Siskiyou region of northern California/southern Oregon is one area identified as containing suitable wolf habitat. The World Wildlife Fund (WWF) classifies this land of ancient redwoods and crystal clear rivers as one of the seven most important ecoregions in the United States. It is one of the top three temperate conifer forests in terms of biodiversity. Brian Barr, program officer for wildlands restoration for WWF's Klamath-Siskiyou region, says that "the wolf is one of the things missing in this ecoregion. Repatriating or reintroducing them is desirable so they can exert their ecological benefit on the rest of the species that still exist here." A 1999 study showed that the region could hold up to 440 wolves. Barr says, “Certainly if gray wolf recovery is considered complete because we have populations in Montana, Idaho and Wyoming, that’s not going to help us out here in California.”

Maine, New Hampshire, Vermont and upstate New York could provide a home for as many as 1,800 wolves, according to recent scientific studies by respected carnivore researchers Dan Harrison and Ted Chapin at the University of Maine and David Mladenoff and Ted Sickley at the University of Wisconsin–Madison. The forests of the Northeast are expanding after generations of agricultural use and deforestation in the 1700s and 1800s. Beavers, once scarce in New York, are once again abundant. Moose have rebounded throughout the northeastern states, particularly in Maine. River otters are being restored in New York. Lynx, long thought absent from this region, have been rediscovered in Maine. Yet the wolf, the region’s top carnivore, is still missing, and scientists believe the species cannot return on its own without an active reintroduction project. FWS indicated in its proposed wolf rule, released in July 2000, that it would pursue wolf recovery in the northeastern United States, but the final rule abandons those efforts.

Colorado’s wolves were extirpated by 1945, but a 1994 feasibility study sponsored by FWS itself showed that the state can still support up to 1,128 wolves. Without them, the willow and aspen on elks’ winter foraging range in places like Rocky Mountain National Park and other public lands will continue to decline. Natural recovery is unlikely, because the Red Desert of southern Wyoming serves as a significant barrier to potential wolf dispersal from existing populations in the northern Rockies. FWS’s rule instead divides Colorado in half, lumping the northern portion as the West and the southern portion as the Southwest. Defenders petitioned FWS in 2000 to develop a recovery plan specifically for the southern Rockies region, primarily Colorado. Because the new rule does not focus on Colorado as a recovery area, chances of attaining complete wolf restoration in this state are remote.

The widespread persecution of wolves that occurred for much of our nation’s history did not cease until people began to better understand the important role wolves play in a healthy ecosystem. Now it is becoming apparent that wolves can even be economically beneficial. The Yellowstone National Park area is estimated to glean millions of dollars annually in increased tourism dollars as a result of people visiting specifically to see wolves. Visitation to Algonquin Park in eastern Canada increases substantially every August when people come from far and wide for the weekly "wolf howls," during which park naturalists imitate wolf howls and often get a response. Ely, Minnesota, home to the International Wolf Center, brings in $3 million annually from this enterprise.

Most people believe there are moral and ethical reasons to restore the wolf as well. Poll after poll across the United States indicates that the majority of the public supports wolf restoration. FWS is the agency charged with restoring our nation's imperiled species, and it has the expertise and resources to make wolf restoration a success. From the Great North Woods of the Northeast to the coniferous forests of the Pacific Northwest, we owe it to future generations to restore our ecosystems and pass on a healthy planet, complete with wolves and quaking aspen.

# Are Protection Efforts Working for Endangered Species Around the World?

*Keystone*, *umbrella*, and *flagship*—these are words conservationists use to describe the role a particular species plays in its ecosystem. A *keystone* species is one whose presence contributes to the diversity and health of the ecosystem. For example, top predators (like wolves) keep the populations of prey (like deer) healthy. An *umbrella* species is one, like the grizzly bear or tiger, that requires a large habitat. When it is protected, many other species benefit from the "umbrella" of its protection. A *flagship* species is chosen because of its appeal to the public for demonstrating conservation issues. The giant panda, whose image appears on the World Wildlife Fund's logo, is an example of this. Some species, like wolves, may even fit into all three categories.

The following articles highlight some of the flagship species chosen by the World Wildlife Fund (WWF) to bring attention to important conservation issues around the world. They include the giant panda, elephants (both African and Asian), the great apes, rhinos, whales, tigers, and marine (sea) turtles. Some good news is that a 2004 survey showed that there are nearly 1,600 pandas living in the wild, over 40% more than scientists previously thought existed. The populations of southern white rhinos and Indian rhinos are growing on protected reserves. A new study shows that marine turtles are worth more alive than dead. However, not enough progress has been made: The world has lost over 90% of its tiger population.

The WWF was founded in 1961, with a mission "to stop the degradation of the planet's natural environment and to build a future in which humans live in harmony with nature, by: conserving the world's biological diversity; ensuring that the use of renewable natural resources is sustainable: and, promoting the reduction of pollution and wasteful consumption."

—The Editor

# Flagship Species Factsheets

from the World Wildlife Fund

The giant panda is universally loved, and of course has a special significance for WWF as it has been the organization's symbol since it was formed in 1961.

Today, the giant panda's future remains uncertain, however. This peaceful, bamboo-eating member of the bear family faces a number of threats. Its forest habitat, in the mountainous areas of southwestern China, is fragmented and giant panda populations are small and isolated from each other. Meanwhile, poaching remains an ever-present threat.

## 30 RESERVES CREATED

The government of China has proclaimed more than thirty reserves for giant pandas, but habitat destruction continues to pose a threat to the many pandas living outside these areas, and poaching is a further problem. Today, around 61% of the population, or 986 pandas, are under protection in reserves.

As China's economy continues its rapid development, it is more important than ever to ensure the giant panda's survival.

WWF has been active in giant panda conservation since 1981, when it supported U.S. scientist Dr George Schaller and his Chinese colleagues in field studies in the Wolong Reserve.

More recently, WWF has been aiding the government of China to undertake its National Conservation Programme for the giant panda and its habitat. This programme has made significant progress: there are now 33 giant panda reserves, protecting over 16,000 $km^2$ [6,177 square miles] of forest in and around giant panda habitat. The latest survey in 2004 revealed there are 1,600 individuals in the wild.

## GIANT PANDA: THREATS

The recent 2004 survey pinpointed a number of threats to the long-term survival of this endangered species, including deforestation and continued poaching.

### Habitat Loss and Fragmentation

Habitat loss due to explosive population growth and unsustainable use of natural resources has pandas clinging to survival across their range, as large areas of natural forest have been cleared for agriculture, timber and fuelwood.

Because of China's dense human population, many panda populations are isolated in narrow belts of bamboo no more than 1,000–1,200 metres [1,093–1,312 yards] in width. Panda habitat is continuing to disappear as settlers push higher up the mountain slopes.

Across the panda's range, habitat is fragmented into more than 20 isolated patches. Within these patches, a network of nature reserves provides protection for more than half of the panda population. Because pandas cannot migrate between these far-flung habitat blocks, they have less flexibility to find new feeding areas during periodic bamboo die-off episodes (see below).

### Illegal Trade

Poaching of pandas still occurs. Even at low levels, this activity can have grave consequences for such an endangered species.

In recent years, several panda pelts being sold for large sums have been confiscated, but there is little information about the dynamics and dimensions of this market. Pandas are also unintentionally injured or killed in traps and snares set for other animals, such as musk deer and black bears.

### Bamboo "Die Back"

In the late 1970s and early 1980s, a number of giant pandas starved to death in the Minshan and Qionglai Mountains following the flowering and die-back of bamboo over wide areas.

Bamboo die-back is a natural phenomenon, occurring every 15–120 years according to the species. Once the bamboo dies it can take a year to regenerate from seed and as long as 20 years before a new crop can support a giant panda population. Bamboo die-back may have helped to disperse giant pandas in the distant past, as individuals migrated to seek areas with other species of bamboo, but now human settlements form a barrier against giant panda movements.

## AFRICAN ELEPHANTS: THREATS

Historically, two major factors have led to the decline of the African elephants: demand for ivory and changes in land-use. Elephants are still poached for ivory and meat in many parts of their range. Remaining elephant habitat is increasingly encroached upon by human settlement and agriculture, leaving many populations restricted to isolated protected areas.

While the illegal trade in ivory remains a real threat, current concern for the survival of the African elephant centres around the reduction of their habitat.

### Habitat Loss and Fragmentation

Most elephant range still extends outside protected areas, and the rapid growth of human populations and the extension of agriculture into rangelands and forests formerly considered unsuitable for farming mean that large areas are now permanently off-limits for elephants.

As habitats contract and human populations expand, people and elephants are increasingly coming into contact with each other. Where farms border elephant habitat or cross elephant migration corridors, damage to crops and villages can become commonplace, providing a source of conflict which the elephants invariably lose.

Inevitably, loss of life sometimes occurs on both sides, as people get trampled while trying to protect their livelihood, and "problem" elephants get shot by game guards. It is predicted that as human populations continue to grow throughout the elephants' range, habitat loss and degradation will become the major threats to elephants survival.

### Illegal Hunting and Trade

In the early 1970s, demand for ivory soared and the amount of ivory leaving Africa rose to levels not seen since the start of the

[20th] century. Most of the ivory leaving Africa was taken illegally and over 80% of all the raw ivory traded came from poached elephants.

This illegal trade was largely responsible for reducing the African elephant population to current levels. The poaching was generally well-organized and difficult to control because of the availability of automatic weapons.

Although international trade in ivory is illegal (except under specific circumstances tightly controlled by CITES—the Convention on International Trade in Endangered Species of Wild Fauna and Flora), there are still some thriving but unmonitored domestic ivory markets in a number of states, some of which have few elephants of their own remaining. These markets fuel an illegal international trade.

### Introduction of the Ivory Ban

As the "ivory ban" came into force in 1990, some countries in Africa experienced a steep decline in illegal killing, especially where elephants were adequately protected. However, in countries where wildlife management authorities are chronically under-funded, poaching still appears to be a chronic, significant problem.

Moreover, increasing land use pressures on elephant range, declining law enforcement budgets, and continuing poaching pressure for bushmeat [meat from native animals, often eaten by local people] as well as ivory, have kept illegal killing of elephants widespread in some regions.

### Unequal Distribution of Elephants Means Different Opinions on Conservation

Considerable debate surrounds elephant conservation, largely because of the varying status of elephant populations in different range countries.

Some people, mainly in southern African countries where elephant populations are increasing, consider that a legal and controlled ivory trade could bring substantial economic benefits to Africa without jeopardizing the conservation of the species, others are opposed to it because corruption and lack of law enforcement in some countries would make it difficult to control the trade.

## CHIMPANZEES: THREATS

The main threats to the chimpanzee are habitat loss and hunting for bushmeat. The relative severity of these threats differs from region to region, but the two are linked. Many conservationists believe that the bushmeat trade is now the greatest threat to forest biodiversity in West and Central Africa.

### Habitat Loss and Degradation

Degradation of forests through logging, mining, farming, and other forms of land development is contributing to the decline of primate species throughout tropical Africa. Remaining habitat patches are often small and unconnected, leaving chimpanzee populations isolated. Deforestation is most advanced in West Africa, where only remnant tracts of primary rainforest remain. The small populations of western, Nigerian, and eastern chimpanzees are primarily located in remnant forest reserves and national parks. In many such "protected areas," poaching for meat and live infants is common, as is unauthorized logging, mining and farming. Logging activities improve access to formerly remote forest areas, leading to increased hunting pressure.

### Bushmeat

"Bushmeat" has always been a primary source of dietary protein in Central and West African countries. In recent years, hunting for bushmeat, once a subsistence activity, has become heavily commercialized and much of the meat goes to urban residents who can afford to pay premium prices for it.

The effect of the bushmeat trade on chimp populations has yet to be evaluated, but a study in Congo showed that offtake was 5 to 7%, surpassing annual population increase. In addition, apes are often injured or killed in snares set for other animals. Infant chimpanzees are frequently taken alive and sold in the cities as pets.

## GORILLAS: THREATS

Gorillas are sought after as food, pets, and their body parts are used in medicine and as magical charms. Besides being in demand for meat, there is widespread belief that gorilla body parts have medicinal or magical properties.

### Habitat Loss

Habitat loss is a major threat to gorillas as forests are rapidly being destroyed both by commercial logging interests and for subsistence agriculture. There is a strong link between habitat loss and the bushmeat trade, as forests opened up by timber companies are more easily accessible to hunters, who often sell meat to employees of the logging companies.

### Bushmeat

The commercial trade in bushmeat, which occurs throughout west and central Africa, may now be more of a threat to African primates than habitat loss and degradation, but the number of gorillas killed annually is unknown. Estimating numbers of gorillas poached is difficult because they are often butchered and eaten on the spot, or their meat is smoked for later sale in towns. Although the great apes constitute only a small proportion of the total species killed for the bushmeat trade, they present easy targets for hunters, and in some areas, such as eastern Cameroon, gorillas are favoured by hunters because of the weight of saleable meat. Gorillas are frequently maimed or killed throughout their range by traps and snares intended for other forest animals such as antelope.

### Disease

#### *The Ebola Crisis*

In late 2002 an outbreak of Ebola hemorrhagic fever was reported in the north of the Republic of Congo on the border with Gabon. By late February 2003, 80 cases [had] been reported in northern Congo, including 64 deaths. Most of the cases (72 ill with 59 deaths) [were] in the district of Kéllé. The human infections coincided with a large-scale die-off of great apes in the region.

In Central Africa, in the area currently affected by Ebola, two great apes exist: the Western lowland gorilla (*Gorilla gorilla gorilla*) and the Central chimpanzee (*Pan troglodytes troglodytes*). Before the current crisis, up to 110,000 Western lowland gorillas and between 47,000 and 78,000 Central chimpanzees were thought to remain. In some areas more than 90% of the population of Western lowland gorilla and Central chimpanzee have been killed. For example, in Lossi Gorilla Sanctuary in Congo, in late 2002, 136 out of 143 gorillas disappeared apparently as a result of Ebola. The disease continues to spread and is now reported in Odzala National Park, a site known to have the highest density of great apes in Africa.

## TIGERS: THREATS

Throughout their range in Asia (including the Russian Far East) tiger populations are threatened, either directly from poaching, or from habitat and prey loss. In many places, they struggle for survival with burgeoning human populations competing for similar resources of food and shelter.

Hunted for their pelt and bones, tiger populations in many areas are dwindling. Until the 1930s, hunting for sport was probably the main cause of decline in tiger populations. Between 1940 and the late 1980s, the greatest threat was loss of habitat due to human population expansion and activities such as logging.

Although sport hunting of tigers has drastically declined, poaching for illegal trade continues to be a major threat. Habitat destruction and decimation of prey populations also contribute to the decline of tiger populations.

Threats to tigers can be separated into 2 categories: poaching and retributive killing, which includes the illegal trade of tiger parts and human wildlife conflict, and habitat destruction and fragmentation, including illegal logging and commercial plantations.

Many range countries lack the capacity and resources to properly monitor tiger and prey populations. Policies conducive to ensuring long-term survival of the tiger are often lacking. Where they do exist, implementation is often ineffective.

### Habitat Loss and Fragmentation

Habitat destruction reduces both tigers and its prey. As a result tigers move into settled areas in search of food, where they are more likely to get killed.

### Illegal Trade

#### *Traditional "Cures" a Curse for Tigers*

In recent years, the illegal hunting of tigers for body parts used in traditional Chinese medicines has become a major problem. The growing prosperity of the Southeast Asian and East Asian economies since the 1970s has led to an ever-increasing demand for these medicines. There are also significant markets amongst Chinese communities in North America and Europe. In India many hundreds of Bengal tigers are known to have been killed by poachers, but this is probably the tip of an iceberg, since most poaching is by definition clandestine and difficult to detect.

### From 70 to 20 Poaching Cases

In Russia, the economic crisis, combined with a relaxation of border controls and a ready access to the wildlife markets of east Asia, led to a dramatic increase in the level of poaching. In the early 1990s, between 60 and 70 tigers were killed every year.

However, the establishment of anti-poaching patrols by the government in 1993 and 1994 with support from several conservation groups, including WWF, has paid off. There were fewer than 20 known cases of poaching in 1995 and 1996. In early 2000, 18 traps were removed by rangers in Primorski Krai in the Russian Far East, according to WWF–Russia.

Today, wild tigers occur mostly in small "island" populations. Such isolated populations are predisposed to inbreeding and are increasingly vulnerable to the pressures of encroachment and poaching. Keeping tiger "islands" intact amid some of the most densely human-populated countries on earth is possible, but offers little hope for the tiger's genetic vigour and long-term viability.

## MARINE TURTLES: THREATS

### Habitat Loss and Degradation

Uncontrolled development has led directly to the destruction of critically important marine turtle nesting beaches. Lights from roads and buildings attract hatchlings and disorient them away from the sea. Instead of finding the ocean, the hatchlings fall prey to predators or die the following day from the heat of the sun. Furthermore, vehicle traffic on beaches compacts the sand and makes it impossible for female turtles to dig nests.

Sea walls and jetties change long-shore drift patterns and can cause erosion or destruction of entire beach sections. Beach restoration projects aimed at protecting seaside buildings, through dredging and sand filling continue to destroy important nearshore feeding grounds and alter nesting beaches.

Additionally, important marine turtle feeding habitats such as coral reefs and seagrass beds are continuously being damaged or entirely destroyed as a result of sedimentation, nutrient run-off from the land, insensitive tourist development, destructive fishing techniques and climate change.

### Directed Take

Hunting and egg collection for consumption are major causes of the drastic decline in marine turtle populations around the world. Green turtles are caught for their meat, eggs and calipee. Calipee is the green body fat that has given the turtle its name and it is the main ingredient in turtle soup.

Researchers estimate that each year poachers take 30,000 green turtles in Baja California and that more than 50,000 marine turtles are killed in Southeast Asia and the South Pacific. Leatherback turtles are killed in some places for their meat and ovaries although in most countries only their eggs are consumed. Olive ridley turtles have been pursued for eggs and their skin used for leather production. In the 1960s, over one million olive ridley turtles were butchered each year on

Mexico's Pacific coast. In many countries, juvenile marine turtles are caught, stuffed and sold as curios to tourists. Marine turtle eggs are considered an aphrodisiac in some countries and eaten raw or sold as snacks in bars and restaurants. In 1996, Mexican authorities seized a truck containing 500,000 olive ridley eggs collected illegally from an important marine turtle rookery.

### Trade

International trade in products such as tortoiseshell from hawksbill turtles, green turtle calipee and leather from olive ridley turtles has exacerbated the quantity of directed take of marine turtles.

During colonial times, European countries were the major importers of marine turtle products. Over the past decades, Japan has emerged as the principal country buying shell (known as Bekko) from tropical countries to produce costly handicrafts. All marine turtle species are currently listed on Appendix I of CITES (the Convention on International Trade in Endangered Species of Wild Fauna and Flora) which prohibits any commercial trade by more than the 166 signatory countries. Even so, trade between non-signatory countries and illegal trade persist.

### Indirect Take

Each year, tens of thousands of olive ridley, Kemp's ridley, loggerhead green and leatherback turtles are trapped in shrimping operations. Marine turtles are reptiles and have lungs so when they cannot reach the surface to breathe, they drown.

Gill nets and long-line fisheries are also principal causes of marine turtle mortality. Worldwide, hundreds of thousands of marine turtles are caught annually in trawls, on long-line hooks and in fishing nets.

### Climate Change

Changing climate and global warming have the potential to seriously impact marine turtle populations. Marine turtles have temperature-dependent sex determination, meaning that an increase in global temperatures could change the proportion of female and male turtle hatchlings and could result in marine turtle populations becoming unstable.

A higher frequency of tropical storms caused by climate change will result in increased nest loss.

Similarly, rising sea water levels threaten to wash out or erode entire nesting beaches. Key marine turtle habitats such as coral reefs are particularly vulnerable to increases in sea temperature. An increase of only a couple of degrees in water temperature is enough to cause bleaching that kills corals and threatens the foundation on which entire communities of species such as marine turtles depend. The 1998 coral bleaching incident is the worst on record, with reports of degradation from countries across all of the world's oceans.

Other stresses such as pollution and sedimentation are likely to compound this threat. Seagrass beds are also increasingly impacted by climate change. Fewer coral reefs and seagrass beds mean less food and refuge for marine turtles so their populations will decline further.

### Pollution

Marine turtles can mistake floating plastic materials for jellyfish and choke to death when they try to eat them. Discarded fishing gear entangles marine turtles and can drown or render a turtle unable to feed or swim. Rubbish on beaches can trap hatchlings and prevent them from reaching the ocean. Oil spills can poison marine turtles of all ages.

### Disease

Many types of diseases have been observed in marine turtles. Recent reports of a rise in the occurrence of fibropapillomas, a tumorous disease that can kill marine turtles, is of great concern. It has been suggested that the increased occurrence may be the result of run-off from land or marine pollution that may weaken the turtles' immune system, rendering them more susceptible to infection by the herpes-like virus that is thought to cause the disease. On some of the Hawaiian Islands, almost 70% of stranded green turtles are affected by fibropapillomas.

### Natural Predators

Marine turtles can lay more than 150 eggs per clutch, and lay several times each season, to make up for the high mortality that prevents most marine turtles from reaching maturity. The subtle balance between marine turtles and their predators can be tipped against turtle survival when new predators are introduced or if natural predators suddenly increase in number as a result of human interference. On nesting beaches in the Guianas, predation by dogs now represents a major threat to marine turtle eggs and hatchlings. In the south-eastern United States, household garbage has become a source of food for raccoons. This has led to a major increase in raccoon populations and the results for marine turtles have been devastating. On some beaches, raccoons now dig up and destroy as many as 96% of loggerhead turtle nests.

# How Do Conservation Organizations Work With Local People, Businesses, and Governments to Protect Land and the Plants and Animals That Live There?

Today, some native people are taking advantage of growing markets in nearby cities and all over the world. There are many economic incentives to take and sell all that can be harvested from the land. This includes cutting down tropical hardwoods, selling animals into the multibillion-dollar illegal animal trade, and killing animals to sell for food called "bushmeat." But after the forests are logged and the species are hunted to extinction, the money stops, and only degraded land remains.

To address this growing problem, the World Wildlife Fund (WWF) has worked for over 40 years to create "a global network of well-managed protected areas, sustaining biodiversity and natural resources across entire ecosystems, helping to reduce poverty . . . and giving space for both wildlife and people."

The following is a section of the WWF's 2003 report, *Conserving Nature: Partnering With People*. It outlines the successes and challenges of various WWF projects around the world. WWF partners with governments, industry, various non-governmental agencies, local communities, and indigenous (local) people to protect land, fresh water, and marine areas. Together, the groups work to develop strategies to restore, protect, and manage important areas in different ecosystems.

But many problems still exist. Certain species are becoming threatened and others are going extinct. Some protected areas are being poorly managed. Some habitats are not even considered part of the protected areas, including the open seas and many types of wetlands. In addition, there are the broader threats of the humans taking water out of the natural system and the effects of global climate change.

—The Editor

## Conserving Nature: Partnering With People—World Wildlife Fund's Global Work on Protected Area Networks

from the World Wildlife Fund

The creation of the world's 100,000 protected areas over the past 130 years represents the largest conscious land-use change in history. Protected areas now cover some 12 percent of the Earth's land surface more than India and China put together.

As the number of protected areas has grown, so too has our understanding of the benefits they provide. In addition to their primary function of preserving biological diversity, we now recognize that they also perform essential environmental services; maintain natural resources; shelter local cultures and spiritual sites; mitigate long-term global threats; reduce border tensions; help reduce poverty through providing livelihoods; and provide economic benefits. However, the world's natural areas remain under threat with disastrous consequences. Plant and animal species are becoming extinct faster than at any time in our history. Fisheries are collapsing, while forests and freshwater systems continue to deteriorate.

The reasons for this are varied. Some protected areas are not well managed. Many habitats are poorly represented in the current network of protected areas, including open seas and coastal areas, marshes and swamps, mangroves, grasslands, and temperate forests. In addition, there are several global threats to protected areas, including climate change, continued conversion of natural habitat, diversion of water from rivers and other freshwater systems, and overfishing.

WWF is working harder than ever to establish a global network of ecologically representative and effectively managed land, freshwater, and marine protected areas. With 40 years experience, targeted conservation goals, and projects combing practical field implementation with high-level policy work in over 100 countries, we are uniquely placed to lead protected area work into the 21$^{st}$ century.

### BUT WE CANNOT DO THIS ALONE

Our partners from indigenous people, local communities, park managers, and NGOs [nongovernmental organizations], to governments, international organizations, development agencies, landowners, and industry—have always been integral to our protected area work. Indeed one of our guiding principles is to build and strengthen working relationships for conservation. Only by working together can we secure the future of our planet s natural areas: bringing benefits to both people and nature.

### CONSERVATION STRATEGIES

As isolated pockets, protected areas alone cannot provide effective conservation of biodiversity. Nor can they maintain viable population of species in blocks of natural habitat large enough to be resilient to large-scale disturbances such as climate change, or provide essential ecosystem services. A key element of WWF's work is therefore conservation of ecoregions.

WWF has prioritized 238 land, freshwater, and marine ecoregions of critical importance to biodiversity—The Global 200 Ecoregions—that are the main areas in which our conservation work is focused. These include diverse land-, water-, and seascapes, including deserts, forests, grasslands, tundra, corals, mangroves, marshes, rivers, and seas.

We are developing large-scale conservation strategies for a subset of these ecoregions in order to complete a representative network of protected areas to safeguard our planet's terrestrial, freshwater, and marine biodiversity. These strategies combine protection, good management, and restoration of landscapes. One important element of this work is to identify "gaps"—areas that are not yet protected but should be for effective protection of the entire ecoregion. In addition, ecosystems that are currently underrepresented in existing protected area networks—such as wetlands, marine and coastal areas, savannahs and grasslands, and temperate forests—need to be included in these networks.

We see national protected areas as just one element—trans-boundary protected areas, buffer zones, corridors, no-take zones, and sustainable land uses within and outside protected areas are also parts of ecoregion conservation. We also consider social and economic factors, both to integrate conservation with sustainable development and to help tackle global issues like the causes and impacts of climate change.

## CONSERVATION

With so much to be done, it is important to focus conservation work to maximize our effectiveness. In order to increase the coverage of protected areas, improve their management, and safeguard them from long-term global threats, WWF has developed conservation programmes with ten-year targets and intermediate milestones by which to measure progress and success. Protected area targets typically address three major issues: the creation of new protected areas; improved management and funding; and the reduction of key threats.

This targeted approach, focused primarily in the Global 200 ecoregions, is paying off. Since the mid-1990s we have helped to create nearly 80 million hectares [198 million acres] of new protected areas, conserving critical forests, freshwater, marine, and coastal environments. A snapshot of our different protected area targets and progress-so-far follows.

### Forests

Since 1998, when the target was first announced, WWF has helped to identify and protect over 30 million hectares [74 million acres] of the world's most exceptional forests. We are currently working to help governments to meet their commitments to set aside an additional 30 million hectares. Also, through strong partnerships, we are currently helping to improve the management of over 350 forest protected areas worldwide, with a combined area of over 40 million hectares [99 million acres] ten times the size of Switzerland. A system for monitoring management improvements has also been put in place. Current work involves completing gap and threat analyses for focal forest ecoregions and mapping target sites to complete representative protected area networks. We are also addressing the need for restoration and design of natural corridors and the curbing of key threats such as illegal logging, forest conversion, forest fires, and the negative impact of major infrastructure development.

### Freshwater

Since 1999, WWF has helped protect 38 million hectares [94 million acres] of critically important wetland areas, many of which have been designated as Ramsar Sites the world's largest single protected area network. Nine countries designated 22 million hectares [54 million acres] of Ramsar Sites in the first eight months of 2002 alone, amounting to a quarter of the total area designated in the previous 30 years. These designations will not only help countries to protect their freshwater sources, but also further their sustainable development goals and address poverty and livelihood issues. With the diversion of fresh water for agriculture soaring around the world, WWF is also working to ensure that the ecology of wetland protected areas is maintained through adequate allocations of fresh water by river basin management programmes.

### Marine

WWF has helped make oceans the new protected area frontier, using our vision, field experience, and science and policy resources to lead the global effort for increased marine protection. During the last few years, as governments have finally begun to recognize the importance of marine protected area networks, WWF has targeted helped achieve protection for more than 10 million hectares [25 million acres] of marine areas, including important coral reefs, sea grasses, fishing zones, and deep-sea habitats. In addition, WWF has worked with local communities to establish no-take zones to help protect fish stocks, and is working for more effective management of marine protected areas.

### Flagship Species

One of the key threats facing the world's species, and therefore biodiversity, is the loss and destruction of important habitats and ecosystems. Using conservation of key species as a focus, WWF is working with governments, scientists, and local communities to safeguard the biodiversity of over 25 key landscapes around the world through identifying key protected areas and training local communities in management techniques. Over the past year, WWF has helped create 2.8 billion hectares [7 billion acres] of ocean sanctuaries important for whales and other species. These efforts will contribute to both the preservation of biodiversity and the economic well-being of many communities through increased opportunities for ecotourism and sustainable activities.

### Climate Change

WWF has developed a *Climate Change Users Manual*, the first-ever tool to help managers of protected areas assess the impacts of climate change and develop strategies to buffer these effects. Covering most of the world's key habitats, the manual gives advice on assessing vulnerability and selecting and devising strategies to enable protected areas to cope with the impacts of limited global warming. Such damage-control strategies must always be implemented in conjunction with efforts to reduce emissions of greenhouse gases in order to keep the increase in average global temperature well below 2°C [35.6°F].

## MAKING POLICY WORK

Protected areas cannot be safeguarded simply by defining a boundary on a map. Long-term conservation needs the support of local, national, regional, and international policy, as well as good governance and good business practice.

### International and National Policy

WWF played a major role in the development of international agreements such as the Convention on Biological Diversity and the Plan of Implementation of the World Summit on Sustainable Development. These agreements call for the establishment of networks of well-managed protected areas.

WWF works with governments, aid agencies, local communities, and others to develop and implement their protected area policies so that these international treaties can become reality on the ground.

We also work with other international agreements that support the role of protected areas. The Convention on International Trade in Endangered Species (CITES) helps to limit trade in threatened plants and animals, thus reducing the incentive to remove these species illegally from protected areas. WWF is working with the Convention on Migratory Species to protect species as they migrate, and with the Ramsar Convention to manage wetlands and coastal marine habitat for the benefit of nature and people.

Trade, development, and other economic issues also affect protected areas. WWF works at the global level to address the links between the environment, trade, poverty, social equity, and macroeconomic reform. We work in close collaboration with governments and communities around the world to ensure that protected areas are well integrated in national policies for development, agriculture, fisheries, and water management.

### Governance

Poor governance typically leads to conflicts over the legitimacy of protected areas and may result in their degradation. WWF advocates and supports the development of improved governance, including appropriate power structures, decision-making processes, and stakeholder involvement to ensure equity, accountability, and performance of protected areas. We also help to build capacity and empower local leadership and grassroots and other nongovernmental organizations to participate and engage professionally in collaborative design and management of protected areas.

### Good Business Practice

Protected areas also need the support of business and industry. WWF encourages practices that prevent the exploration, exploitation, or damage of protected areas and promote the sustainable use of natural resources, whether publicly or privately owned.

## PEOPLE AND PROTECTED AREAS

Indigenous peoples and local communities are crucial for the protection of natural resources and biodiversity. Indigenous peoples, for example, inhabit nearly 20 percent of the planet, mainly in areas where there is still a high degree of biodiversity. This makes them some of the Earth's most important stewards. When their lands come under threat from non-sustainable forms of development, they also suffer. WWF recognizes the crucial role that indigenous people and local communities play in safeguarding biodiversity through the perpetuation of those traditional practices that ensure the sustainable use of natural resources. We believe that protected areas are only viable if they are supported by indigenous peoples and local communities who live in or near the area and depend on it. Our work on protected areas therefore not only aims to safeguard biodiversity but also to sustain the cultures and livelihoods of people.

We respect the human and development rights of indigenous peoples and local communities, and recognize the need to balance biodiversity conservation with peoples livelihoods with no net loss to either. Protected areas have sometimes been established with little regard for indigenous peoples and local communities, who may have been forcibly removed from their traditional lands or lost their rights to the land and access to natural resources, suffering as a result. Building support from local communities and indigenous peoples and securing their improved livelihoods are critical to ensuring effective management of protected areas. If indigenous or local people have been forced off land or lost ownership and access rights in the creation of protected areas, restitution measures should be considered.

WWF believes that involvement of communities in protected areas must start at the planning and establishment phases, be carried through to management and monitoring, and include benefit sharing. In the implementation of its programme,

WWF is strongly committed to identifying the economic, social, and cultural benefits of protected areas to people. We work on strengthening the role of protected areas in providing sustainable livelihoods and food and water security, particularly to the poor. We help ensure that protected areas can continue to deliver essential environmental services to society at large, maintain natural resources, mitigate global long-term threats, and keep safe havens for the expression of cultural and spiritual values. . . .

### REWARDING CONSERVATION EFFORTS

Our partners invest a lot of time, effort, and money to achieve conservation goals. To recognize this, WWF has developed the Gift to the Earth scheme, where governments, companies, and communities are publicly recognized for their contributions to conservation.

The process for public recognition is not usually spontaneous, but is developed through discussion and negotiation with the potential recipient. This helps ensure that new protected areas are set up with scientific credibility, stakeholder involvement, integration into a wider landscape mosaic, and funding.

In partnership with the Alliance of Religions and Conservation (ARC), this mechanism has been extended by WWF to sites of spiritual or religious importance which can now be recognized as Sacred Gifts.

Since the launch of this scheme in 1996, WWF has recognized and celebrated 88 Gifts to the Earth all over the world. These include new marine protected areas in Australia, the Azores, Malaysia, Mozambique, Norway, and the United Arab Emirates; freshwater protected areas in Bolivia, Chad, China, Mongolia, and Zambia; and forest protected areas in Canada, Gabon, Madagascar, Russia, and Turkey.

### CHALLENGES AND FUTURE WORK

If the last century [the 20th century] was the time of protected area creation, the next hundred years must be a time of completing networks, paying for and organizing their management, integrating them into wider society, and protecting them against global threats.

This will be no easy task, and the scale of the challenges should not be underestimated.

Important ecoregions remain underprotected, and many protected areas are isolated and fragmented.

Effective protection remains frustratingly rare. Many protected areas exist in name only, so-called paper parks, or do not have enough resources to handle immediate threats such as illegal activities (for example, fishing, land clearing, mining, and poaching), diversion of water, pollution, uncontrolled tourism, or longer-term problems such as climate change.

In addition, there are worrying signs of government and donor fatigue. Caught up in the immediate problems of economic downturn, civil unrest, or war, government aid agencies have diverted their attention away from protected areas, while foundations and private donors have reduced their contributions.

# What Programs Serve to Protect Our Important Ocean Habitats?

Marine (ocean) ecosystems around the world are threatened by coastal development, pollution from many sources, invasive (introduced) species, overfishing, and even global warming. Sometimes, when one species disappears from an area, it can affect the entire ecosystem—the organisms in an area and their environment. For example, when the population of sea otters along California's coast declined, the population of sea urchins, their usual diet, grew. The sea urchins then ate so much of the kelp (seaweed) forests, that the fisheries in the area were damaged. When the sea otters were protected, they again ate more urchins, which allowed the kelp to grow back. The fish once again had a place to live and spawn.

Some examples of marine ecosystems include the deep ocean, coral reefs, estuaries, and continental shelves. These ocean resources have huge economic and environmental value. But unlike pollution or problems of misuse on the land, damage to the oceans has been easier to overlook.

As people became more aware of the ocean-related issues, in 1972, the United States passed the Marine Sanctuaries Act.[1] In marine protected areas (MPAs), human activities are regulated, but not necessarily prohibited. Since 1972, the United States has designated 13 national marine sanctuaries. It is within these protected areas that smaller areas called marine reserves can be designated. In marine reserves, no resource can be extracted and no habitat can be destroyed. Yet marine reserves, unlike marine protected areas, tend to be small. Most are less than 0.5 square miles [1.3 km$^2$].

The following section is from a 2003 report prepared for the Pew Oceans Commission. In June 2000, 18 members of the independent Pew Oceans Commission began the first review of U.S. ocean policy in over 30 years. The members of the commission came from all areas: fishing, science, conservation, government, education, and business. They traveled the country talking to everyone with connections to the oceans, from fishermen to farmers. They made their formal recommendations to Congress in June 2003. As noted in the report's conclusion, marine reserves have the potential to preserve and protect marine ecosystems because of their connection to larger systems. It is becoming clear that saving these smaller pockets has significant and far-reaching value.

—The Editor

1. National Marine Sanctuaries NOAA Factsheet. *Welcome to the National Program*. August 2002. Available online at *http://www.sanctuaries.nos.noaa.gov/natprogram/natprogram.html*.

## Marine Reserves: A Tool for Ecosystem Management and Conservation

from the Pew Oceans Commission

### AN INTRODUCTION TO MARINE RESERVES

The ocean is a global highway, a self-filling pantry, and the Earth's lungs. Its influence rises far above high tide, washing into the lives of every human—even those with homes in communities far inland. However, it isn't the sea itself that does yeoman's duty in supporting human populations—it is the life of the sea. Without the sea's microbial plant and animal species to produce oxygen, absorb $CO_2$, produce food, break down wastes, stabilize coastlines, and aerate sediments, life above sea level would be greatly different. For centuries, the vastness of the sea and the bounty of its life made any human-induced deterioration nearly unthinkable. "I believe, then, . . . that probably all the great sea fisheries, are inexhaustible . . ." reassured nineteenth-century biologist Thomas Huxley.

Unfortunately, they are not. Entire marine ecosystems are affected at nearly every level by a variety of threats—from overfishing to chemical pollution and physical alterations. The ability of ecosystems to absorb these impacts is pivotal to the long-term health of the oceans. Evaluating and responding to these threats in an integrated fashion is the most critical management challenge.

The goal of this report is to summarize current information on one emerging tool in marine ecosystem management—fully protected marine reserves. To place this tool in its biological perspective requires an additional focus on marine ecosystems, how they function, the types of threats they face, and how reserves can address them. Topics such as the mechanisms that governments have used to establish reserves and new social and economic approaches to evaluating them are introduced but comprehensive treatment of these is outside the scope of this report. The following chapters focus on the biology of reserves, how they fit into current management schemes, how they differ from traditional management, and the hope they bring to solving daunting problems. The diversity of ways reserves can function in marine management is one of their principal virtues. However, some management goals cannot be reached with reserves, and these need to be noted as well.

The scientific evidence on the benefits of reserves is overwhelming ion some areas, currently emerging in others, and frustratingly poor at some crucial junctures. An important goal of the report is to highlight these different levels of knowledge in order to foster a sense of the practical utility of reserves in future management of the oceans.

## THE NATURE OF RESERVES

Marine reserves are a special category of marine protected areas. Whereas any marine habitat in which human activity is managed is a marine protected area, marine reserves are areas in which no extractive use of any resource—living, fossil, or mineral—nor any habitat destruction is allowed. These areas are generally called fully protected marine reserves, and they represent the major management tool discussed in this report. Other, less comprehensive levels of protection from extraction—seasonal closures, bans on taking reproductive individuals, and catch limits—are common in U.S. marine habitats. There are also areas in which mineral extraction or waste disposal is restricted. Any area in which these types of habitat or species protection occur can be called a marine protected area.

Because there are so many different types of marine protected areas, a map of their placement can be confusing or even misleading. Large areas of seemingly protected coast may, in fact, provide poor comprehensive protection. For example, the large expanses of the 13 national marine sanctuaries seem to be the crown jewel of the U.S. marine reserve system. However, these sanctuaries provide protection mostly against oil and gas development. Fully protected marine reserves only exist where they have been carefully negotiated with the local community.

Even large numbers of marine protected areas may include few reserves: Marine protected areas in California number over 100, but less than a quarter of one percent of their combined area is completely protected from fishing. In the Gulf of Maine, there is an impressive mosaic of protected areas, but full protection is implemented in only three tiny wildlife refuges. In most areas, protection is limited to a single species or is focused on a single activity such as oil exploration.

## WHERE ARE MARINE RESERVES?

The scientific and marine management literature sports abundant examples of fully protected marine reserves—hereafter, simply called marine reserves—established along many different coastlines around the world. Over the last 30 years, reserves have been established along coral reefs, temperate shores, in estuaries, mangroves, and many other habitats. Despite these examples, the area protected in marine reserves is still a tiny fraction of one percent of the world's oceans, a small figure compared to the four percent of global land area protected in terrestrial parks. Across North America, the area protected in state and federal parks outstrips the area in marine reserves by a ratio of 100 to 1. Marine reserves also tend to be small—the vast majority are one square kilometer (0.39 $mi^2$) or less in area. Brackett's Landing Shoreline Sanctuary Conservation Area (formerly named Edmonds Underwater Park) in Puget Sound, Washington, one of the oldest marine reserves in the

U.S., contains just 0.04 square miles (0.10 $km^2$). By contrast, the De Hoop reserve in South Africa spans about 150 square miles (388.4 $km^2$).

Marine reserves in the U.S. are limited to a thinly scattered set of research and recreational sites, such as the Big Creek Marine Resources Protection Act Ecological Reserve, in California, or the Brackett's Landing Shoreline Sanctuary Conservation Area, in Washington. Larger reserves occur in the Florida Keys National Marine Sanctuary, where the Western Sambo Ecological Research Reserve is about 12 square miles (31 $km^2$) of seagrass, coral, and reef. The newly designated Tortugas Ecological Reserve weighs in at 200 square miles (518 $km^2$) and is the largest in the U.S. Although a comprehensive list of U.S. marine reserves does not yet exist, a review of the literature and current Websites turns up only about two dozen fully protected marine reserves.

## AUTHORITY TO CREATE RESERVES

The authority to create marine reserves remains unclear for the vast majority of U.S. ocean waters. The National Marine Sanctuary Program provides a process for establishing reserves within a sanctuary boundary, with implementation and enforcement through existing state and federal agencies. The regional fishery management councils can restrict removal of species within their control, but they cannot set aside an area as a closure for all species. The lack of clear authority can also be found at the state level. A notable exception is the state of California, which passed the Marine Life Protection Act to provide the governance framework for marine reserves. Other states and the U.S. federal government do not have such a blueprint, making the process for establishment of marine reserves unclear.

The lack of clear authority to establish marine reserves makes their implementation as tools for ecosystem-based management more difficult. This confusion can obscure a critical point agreed on by virtually all proponents of marine reserves—that the public, as stakeholders for marine ecosystems, must be involved at the early stages of any plan to implement reserves. Without clear guidelines for where the authority to establish reserves exists, involving stakeholders becomes more difficult. . . .

## THE ECOSYSTEM CONTEXT OF RESERVES

Marine reserves were designed to reduce the impact of human activity on marine ecosystems, particularly the ecological damage caused by overfishing of some coastal areas. Focus on single fish species is the traditional method of fisheries management. Recent collapses of important U.S. fisheries such as New England cod (*Gadus morhua*) and the struggle of many U.S. fishing communities have prompted a call for evaluation of alternative methods. Because they protect habitats and all the species that use them, marine reserves are a management tool that affects the representative parts of whole ecosystems. The value of this approach comes from the differences between protecting all the species in a functioning ecosystem versus protecting a few species through focused management. A core understanding of why ecosystems are more than just the sum of their parts requires a glimpse into how ecosystems work, and what maintains them. The past few decades have witnessed enormous advances in the understanding of marine ecosystem function, and this timely information can be brought to bear on the issue of marine reserves.

# Is Our National Wildlife Refuge System Working?

Our National Wildlife Refuge system recently celebrated a big birthday. In the early 1900s, President Theodore Roosevelt had the idea that, although we should use our natural resources, we should also be sure to protect them so that there would be plenty remaining for future generations to enjoy. In 1903, Roosevelt set aside a small island in Florida called Pelican Island. He wanted to protect the pelicans, egrets, and roseate spoonbills that were in danger of going extinct because their feathers were being used to adorn ladies' hats. Thus began the National Wildlife Refuge system.

The mission of the National Wildlife Refuge System is "To administer a national network of lands and waters for the conservations, management and where appropriate, restoration of the fish, wildlife and plant resources and their habitats within the United States for the benefit of present and future generations of Americans."

Today, 100 years later, there are more than 540 American refuges, with at least one in every state. In fact, you may live near one. The total protected land area is about the size of the states of New York, Pennsylvania, and Virginia combined: 95 million acres [38 million hectares] that include the habitats of hundreds of endangered or threatened species.

Refuges come in all shapes and sizes. Some are huge areas of land, serving as homes to large mammals. Others link larger habitats together. Today, some refuges are part of critical habitats, as in the prairie pothole region where these small wetlands are crucial as stopovers for migrating waterfowl. Sometimes the ecosystems in refuges have to be restored to natural conditions, as in Hawaii's Hakalau Forest National Wildlife Refuge.

Although Theodore Roosevelt envisioned taking action before resources were squandered, that has not happened according to his plan. Still, there is hope. "The enduring lesson of the whooping crane," says Tom Stehn, the species' recovery coordinator for the

U.S. Fish and Wildlife Service, "is that it is never too late. If man steps in and cares enough and tries hard enough, we can save wildlife species that seem utterly doomed."

The following article by T. Edward Nickens, "The Art of Helping Wildlife," highlights some of the refuge success stories. In one of them, the author himself assisted endangered leatherback turtles on a refuge in the U.S. Virgin Islands. Before the refuge was set up with the work of volunteers, the beach had only 19 female turtles nesting. Since the refuge was established in 1984, more than 6,000 nests have been reported. That means more than 250,000 turtles have hatched.

—The Editor

## The Art of Helping Wildlife

by T. Edward Nickens

One by one, we reach out to the turtle. She's 10 feet from the surf line and gasping for breath, her lungs unaccustomed to the unsupported bulk of her 700-pound frame. "So huge," coos one member of our group, giving the leatherback sea turtle a reassuring pat. Others huddle nearby, measuring the turtle's carapace, taking a blood sample, injecting a tiny computer chip into her shoulder.

I am one of seven volunteers gathered on this beach in the U.S. Virgin Islands to assist the endangered leatherbacks. On my belly behind this laboring turtle, half-buried in the sand, my hands reach into a deep nesting pit to catch the lemon-sized eggs as they drop from her cloaca. I pass the eggs quickly to other helpers. Later this evening we'll rebury them on a more stable part of the beach.

This mama sea turtle has chosen to nest on an erosion-prone section of the beach, and there is little chance her eggs will hatch if they remain here. Her good fortune, however, is that this shoreline is part of the Sandy Point National Wildlife

Refuge, which hosts the largest and most intensely managed colony of nesting leatherback sea turtles in the United States. "Without its protection as a federal refuge, that nesting beach would be a resort and the leather-backs would be in really big trouble," says Donna Dutton, a biologist leading the U.S. Virgin Islands Department of Planning and Natural Resources' leatherback project.

Our presence on this dark Caribbean beach is a testimony to an evolution of thought about how humans can help wild creatures on national wildlife refuges. Scientists have learned that protecting land is often not enough to shore up imperiled species. Active management and hands-on assistance—by scientists and everyday citizens—can be a critical component of wildlife conservation. It's an idea expressed in different ways in different places, and it doesn't always work. But when it does the results can be spectacular: Ducks streaming over marshes; canyons filled with the clattering of battling desert bighorn sheep; and Hawaiian rain forests glittering with brightly colored endangered birds. "The art of wildlife management means we constantly adapt what we learn to the lands we care for," says Jim Kurth, deputy chief for the National Wildlife Refuge System. "That's what really characterizes the way we meet wildlife needs on refuge lands."

On a summer night on the St. Croix beach, you don't have to look far to see humans giving wildlife a helping hand. Time and again biologists and my fellow volunteers reach out to the turtles—to take blood samples for genetic analysis, to gently push hatchlings over driftwood and sea grape. This is a stark contrast to conditions before Sandy Point was established in 1984. Back then, nearly 40 percent of turtle nests were lost to beach erosion, and most of the rest were plundered by local egg poachers. In 1982, only 19 females nested at Sandy Point, and they produced a paltry 2,200 hatchling turtles.

The refuge now hosts a half-dozen teams of volunteers from the Earthwatch Institute each year. This labor allows scientists to tag and monitor every nesting leatherback (the largest of the world's seven species of sea turtles) and analyze every nest.

"This is cutting-edge work," says Dutton. "We now have an incredible amount of information on each turtle that comes to the beach. We know how many times it has nested, how many eggs it has produced, how many times it has returned in each season. Now we're getting genetic fingerprints that are helping us model the population. The protection is critical to the research, and the research is critical to bringing this population back." Since the refuge at Sandy Point was established, some 6,000 turtle nests have been documented on the beach, and nearly a quarter-million leatherback hatchlings have emerged from its sands. In 2001 alone, Sandy Point's beaches drew nearly 10 times as many nesting sea turtles as in 1982, and the refuge produced 44,325 palm-sized hatchlings.

Protecting critical habitat is a cornerstone of the National Wildlife Refuge System and is one aspect of conservation that has seen great change. Consider the so-called "duck factory," the prairie pothole region of the plains states where millions of wetland depressions produce ducks and geese that ultimately migrate across much of the country. Early on, biologists understood the need to protect waterfowl habitat throughout the migratory flyways. But they didn't always understand the need for treating refuges as an integral part of a larger landscape. Those lessons were learned in part at prairie refuges such as the J. Clark Salyer National Wildlife Refuge, along the lower reaches of North Dakota's Souris River. Salyer, a mosaic of marsh, remnant prairie and sandhills, was one of many havens established to give waterfowl a chance to weather the Dust Bowl of the 1930s, when drought turned the prairies into a desert, and drove waterfowl populations to alarming lows.

In its early years, explains refuge manager Bob Howard, Salyer and other waterfowl refuges were managed in isolation. "We bought the land and put a fence around it," he explains, "then worried about what we could do inside the fence." Managers worked to provide cover by planting alfalfa and wheatgrass, and boosted nesting opportunities with chicken-wire nesting baskets and small nesting islands pushed up with bulldozers.

These days, however, the Salyer refuge is managed as just one part of a sprawling, five-county area the size of Connecticut and Rhode Island combined. Here refuge personnel work with private landowners to promote wetlands and native grassland conservation. "Biological and ecological integrity are what we are after," says Howard. "We've learned to look beyond the fences and consider the entire landscape."

Protecting large, intact landscapes is crucial for the survival of many animals, including the four subspecies of desert bighorn sheep. In the mid-nineteenth century, several hundred thousand of these buff-colored cliff-dwellers clattered across dry mountain ranges in the Southwest. Today, perhaps 15,000 remain. More than three million acres of refuge lands have been set aside for desert bighorns, more than has been designated for any other mammal species outside of Alaska.

Preserving untrammeled wilds is only part of the story with desert bighorns, however. Healthy desert bighorn sheep can go many months without drinking water, gleaning what moisture they need from grasses, cacti and shrubs. But during the dry summer months, free-standing water is critical to the herds, so biologists and volunteers at Nevada's Desert National Wildlife Refuge have built water collection and dispensing systems in remote mountain areas. "In many places we've been able to save nearly every drop of rainwater," says Bruce Zeller, a biologist for the Desert refuge. "It's meant hundreds of thousands of gallons of water a year for sheep." It is grueling work: All materials must be ferried in via helicopter or horse, and crews endure difficult weather, tough hikes and days of manual labor.

Desert bighorn sheep still present difficult challenges to biologists. On the Desert refuge in particular, sheep numbers have fallen—from 1,600 in the mid-1980s to a current estimate of approximately 750. Zeller thinks drought and increased numbers of mountain lions may have caused the decline, but biologists still don't have all the answers.

Similar struggles exist across the country and illustrate some challenges—even limitations—of wildlife refuges. For example, flooding on the Julie B. Hansen National Wildlife

Refuge in Washington in 1996 wiped out half of its endangered Columbian white-tailed deer. And scientists are currently trying to learn why the moose population on Minnesota's Agassiz National Wildlife Refuge has plummeted by more than 75 percent in the last two decades. The lesson from these and other setbacks is that wildlife populations go through peaks and valleys, Zeller says, "and we're learning that you have to be patient and let nature take its course. And that whenever you can, you give nature a hand."

On some refuges, giving nature a hand means rebuilding the native ecosystem. At Hawaii's Hakalau Forest National Wildlife Refuge, for example, managers are trying to help rare birds by bringing back native tropical forests. Two hundred years ago this area on the upper slopes of Mauna Kea, the highest peak on Hawaii's "Big Island," was cloaked in a dense canopy of koa and red-blossomed 'ohi'a trees. When the refuge was established in 1985, says manager Richard Wass, much of its 33,000 acres was still heavily wooded. The uppermost 5,000 acres, however, "had been denuded by 150 years of ranching," Wass explains. "A flourishing rain forest had been degraded to a cattle pasture of alien grasses and weeds."

To restore the forest, refuge staff and volunteers have planted more than 250,000 native trees and shrubs on its close-cropped pastures, including 200,000 koa trees, known as the "mother of the rain forest" for the way it nurtures understory vegetation. Forty-four miles of fence have been constructed to keep out grazing cattle and feral pigs, and war has been waged on exotic plants such as gorse, banana poka and blackberry. "If we just let the forest alone, it would probably take 500 or 1,000 years to come back on its own," figures Wass. "But in 50 years, we'll have a diverse forest up there. And in 200 years, it will look almost like it did 200 years ago."

It won't take that long for Hakalau Forest's pioneering restoration to make a difference for wildlife, including some of the nation's most endangered species. Already native birds, including the yellow and black 'amakihi and the brilliant crimson 'apapane, have nested in the regenerating forests. And in

November of 2001, refuge biologist Jack Jeffrey found a family of endangered 'akiapola'au feeding in a grove of planted koa trees. "That's exciting," Jeffrey says. "That shows us we're doing something right. We always said, 'Plant it and they will come,' and they did."

Landscape restoration is becoming a hallmark conservation tool on many wildlife refuges. On the Neal Smith refuge in Iowa, biologists are working to recreate more than 8,000 acres of tallgrass prairie and oak savanna. On the Carolina Sandhills refuge in South Carolina, restoration centers on the longleaf pine-wiregrass ecosystem that once covered 90 million acres of the Southeast. Now found only in scattered patches, this landscape provides critical habitat for endangered red-cockaded woodpeckers and pine barrens tree frogs.

Perhaps no animal symbolizes the helping hand of humans on wildlife refuges as much as the endangered whooping crane. In recent years biologists have snatched the whooping crane from the maw of extinction, and refuges have played a seminal role.

In 1941, only 15 migratory whoopers remained alive in all of North America. (A handful of nonmigratory birds lived in Louisiana, but died out by 1950.) Careful protection of that flock, which migrates between Aransas National Wildlife Refuge in Texas and Wood Buffalo National Park in Canada, has brought its numbers up to about 175 birds. Efforts to establish the birds in the Rocky Mountains failed, but a captive-reared nonmigratory flock of nearly 90 birds has been established in Florida.

And in the fall of 2001, seven whooping cranes completed a 1,230-mile migration flight led by an unusual foster family—ultralight aircraft piloted by men in whooping crane costumes. Based on the pioneering work of Canadian sculptor William Lishman, who led 18 Canada geese from Ontario to Virginia in 1993, researchers are establishing a flock of whoopers that will migrate from Wisconsin to Florida.

Young whooping cranes in this program need to be taught migration routes. Their instruction begins before they hatch, in incubators at the U.S. Geological Survey's Patuxent Wildlife Research Center, where taped crane calls and ultralight aircraft noise play in the background. As soon as chicks hatch, workers draped in white costumes commence their role as caregivers, pointing out mealworm treats with puppet arms mimicking the necks and heads of adult cranes. By the time the birds can fly, they have been moved to Wisconsin's Necedah National Wildlife Refuge to begin daily exercises with another costumed "parent" at the helm of an ultralight aircraft. In October, the plane shepherds the birds south for a six-week migration.

Success has been measured in the trip back to Wisconsin, which the birds undertake by themselves. In the spring of 2002, five whooping cranes retraced the route to Necedah. Officials hope that three times as many will fly back to Wisconsin this spring.

The enduring lesson of the whooping crane, says Tom Stehn, the species' recovery coordinator for the U.S. Fish and Wildlife Service, "is that it is never too late. If man steps in and cares enough and tries hard enough, we can save wildlife species that seem utterly doomed."

And just as important, we can prevent less imperiled animals from ever having to peer into the brink. Back on the beach at Sandy Point refuge, I rebury the eggs laid by the big mama turtle in a stretch of beach safe from erosion, then open up a Styrofoam cooler. It is 3 A.M. and I have the task of releasing tiny leatherback sea turtles that were unable to make it out of nests and into the water on their own. As I pull the hatchlings out of the cooler and place them on the sand, they wriggle urgently in my hand. Waves wash up on the beach like arms of white foam, wrapping around the tiny animals, and then recede into the sea, taking them home. I watch for a few moments, but don't tarry long. There's more work to do.

Section D

# Other Programs for Wildlife Protection

# What Programs Are Protecting the Endangered Orangutans?

When scientists looked at the DNA sequencing, or the genetic information, of humans and great apes, they found that we share over 98% of our genetic information with chimpanzees, and about 96% of the DNA sequencing in all four species of great apes (chimps, bonobos, gorillas, and orangutans) is identical to that of humans. So it should be of concern to all of us that the great apes, our closest genetic relatives, are in trouble—especially since we are responsible for their plight.

Humans have logged and burned the great apes' habitats. We have long hunted them for meat, called bushmeat, but people now hunt apes more to sell in cities and to other markets. People fight destructive wars in their habitats, and we give them diseases. The western chimpanzee is now gone from four of the African countries in which it used to live. The Cross River gorilla is the most threatened of the African apes, with fewer than 250 animals left alive on the planet. If the current rates of loss continue, the only Asian ape, the orangutan, could be extinct by the year 2020.

Yet there is hope. Conservationists and international organizations are working in Africa and Asia to protect the apes' habitats and to save them from hunters. In 2000, President Bill Clinton signed the Great Ape Conservation Act into law. This law authorized spending up to $5 million per year for 5 years to fund projects to help the endangered apes. There are networks of organizations involved in protecting the great apes, including the World Wildlife Fund, the Koko Foundation, and GRASP (the Great Apes Survival Project), a United Nations program.

The following articles examine the work and successes of another international conservation organization, The Nature Conservancy (TNC), in protecting a population of orangutans in Indonesia. Included is a press release explaining the findings of TNC's 2002 report. Also included is a July 2004 press release outlining the agreement reached to more closely monitor the logging in the orangutan's forest habitat, bolstered by a $1 million grant from The Home Depot Company.

—The Editor

## Summary of Orangutan Surveys Conducted in Berau District, East Kalimantan

from The Nature Conservancy

Orangutan populations on both Sumatra and Borneo are in drastic decline. Despite their officially protected status in both Indonesia and Malaysia, it has been estimated that unless the joint threats of deforestation, fragmentation and poaching are seriously confronted in the immediate future, orangutans will be extinct in the wild by the year 2020. A key first step in seriously addressing these threats is to identify all potentially viable orangutan populations and estimate their size. The goal of this study was to assess the abundance of orangutans in an 140,000 hectare area (approximately 346,000 acres) of the Berau District, East Kalimantan. Estimates of orangutan density for this area are provided based on nest counts along line transects. A total survey effort of 71 km [44 miles] of transects was distributed within the area, and yielded an extrapolated total estimate of approximately between 1,000 and 2,500 orangutans in the 140,000 ha [hectare] area. This number represents approximately 10% of the estimated total orangutan population in the world (27,000 in 1997). Preliminary ecological surveys indicate that the survey area is of high quality and relatively low disturbance, and suggest that the area could serve as a major stronghold for the world's rapidly diminishing orangutan population. It is therefore recommended that conservationists consider this location a high priority area in which focused efforts may make a substantial contribution to maintaining populations of wild orangutans. In addition, this result highlights the value of biological surveys of new areas as an important tool in identifying previously unknown but potentially important strongholds for highly threatened species.

Not only is the population in the survey area one of the largest populations left on Borneo, it is also one of the few remaining areas in which focused efforts to maintain a viable population of wild orangutans have a reasonable chance of success. The survey results and subsequent efforts to protect this prime orangutan habitat should therefore be widely publicized and strongly supported by conservationists as one of the last hopes for the protection of wild orangutans from extinction.

### Summary

We assessed the density of orangutans in a 140,000 ha area (approximately 346,000 acres) of unlogged and lightly logged concession forest in the Berau District, East Kalimantan. Our estimates of orangutan density are based on the widely accepted methodology of systematically counting orangutan nests along line transects. We divided the survey area into two different zones—a 50,000 ha zone which has been relatively unexposed to hunting in the recent past (Gunung Gajah) and a 90,000 ha zone in which orangutans have been subject to long-term, intensive hunting (Sungai Gie). A total of 71 km of systematic transects were distributed within the Gunung Gajah zone, and yielded an estimated orangutan density of 2.00 orangutans/km$^2$ (based on an estimated 6.93 nests/ha). An additional 22 km [14 miles] of preliminary transects and a two-week exploratory expedition were conducted in the Sungai Gie zone, indicating that orangutan densities in this portion of the survey area are very low. We conservatively estimate a total population size of between 1,000–2,500 orangutans in the survey area. We also conducted ecological surveys that indicated that the forest there is high quality habitat for orangutans, and that due to the impacts of hunting, the present orangutan population is probably well below the carrying capacity of the area. These results suggest that if this forest is effectively protected from hunting it could serve as a major stronghold for the world's rapidly diminishing orangutan population.

### Rationale

The orangutan (*Pongo pygmaeus*) is Asia's only great ape, and is found only in Indonesia and Malaysia, on the islands of Sumatra and Borneo. Orangutan populations on both islands are in a dramatic state of decline, largely due to hunting, habitat loss and fragmentation. Despite their officially protected status, it has been estimated that unless the joint threats of deforestation, fragmentation, and poaching are seriously confronted in the immediate future, orangutans will be extinct in the wild within the next two decades. The long term protection of wild orangutans will only be achieved if a set of replicated, viable populations can be identified and adequately protected. Although nobody knows how big an orangutan population must be in order to be viable, the basic principles of conservation biology suggest that the most effective strategy is to focus conservation efforts on the few remaining large wild populations. Large populations are at relatively lower extinction risk than are small populations, and their protection is more cost efficient than protecting several smaller, fragmented populations.

The Berau district in East Kalimantan is one of the few remaining areas on either Sumatra or Borneo that still contains substantial expanses of relatively undisturbed lowland Dipterocarp forest. Recent ecological surveys indicated that the Berau District contained an orangutan population that has thus far received little attention from orangutan conservationists.

However the size of this population was unknown. We therefore identified a 140,000 ha (approximately 346,000 acres) block of timber concession forest which was relatively undisturbed, in which local people had reported relatively high orangutan densities, and which was fairly remote from human settlements and therefore less susceptible to land-use conflicts. We conducted orangutan surveys in this area using established nest transect techniques to assess the size of the orangutan population in this area.

## The Nature Conservancy Finds Population of Wild Orangutans

from The Nature Conservancy

Researchers for The Nature Conservancy surveying a remote forest on the Southeast Asian island of Borneo have found a large population of orangutans, the size of which was previously unknown. This find increases the number of known orangutans by approximately 10 percent and offers conservationists a rare, hopeful opportunity in the race to save this highly endangered primate.

Research teams recently documented 1,600 orangutan nests, indicating that between 1,000 and 2,500 orangutans are living within a 540-square-mile [1,399 $km^2$] area of lowland forests in the province of East Kalimantan, part of the Indonesian portion of Borneo. This is the largest viable population of wild orangutans known to exist in East Kalimantan, a province about the size of New England.

"This find represents one of the last, best chances to protect a large, healthy population of wild orangutans anywhere in the world," said Steve McCormick, the Conservancy president and CEO.

Later . . . the Conservancy will sign a joint declaration with the Berau District of East Kalimantan—the district in which the orangutan population is located—and the Indonesian Ministry of Forestry, committing all three parties to conserving and managing orangutan habitat in the district. The document states that the central and local Indonesian governments and the Conservancy recognize the orangutan's importance and the need to conserve its habitat. They also pledge to protect the function and ecology of the orangutan habitat area and agree to promote forest conservation through forest certification.

Experts estimate between 14,000 and 25,000 orangutans are left in the wild, found only on Borneo and Sumatra, another Southeast Asian island. Unless uncontrolled deforestation, habitat fragmentation and poaching are halted immediately, experts believe orangutans will be extinct in the wild by the year 2020.

Indonesia, home to some of the Earth's most biologically diverse and highly threatened tropical forests, is the world's top supplier of wood. It is estimated that 70 percent of the wood taken from Indonesian forests is harvested illegally.

To combat the threats to orangutans posed by unsustainable and illegal logging, the Conservancy is actively working with local East Kalimantan communities and the Indonesian government to create economic incentives to manage forests sustainably and protect prime habitat.

In support of this work, The Home Depot is giving the Conservancy $1 million to be used over the next five years to combat illegal logging and promote sustainable timber harvesting—even though less than one percent of the company's wood supply comes from Indonesia.

"The Home Depot has led the retail industry toward sustainable forestry by using its purchasing dollars to show the company's preference for certified wood," said Ron Jarvis, The Home Depot merchandising vice president for lumber and building materials.

The Home Depot, the world's single largest buyer of wood products, announced in 1999 that it would give purchasing preference to wood certified by the Forest Stewardship Council. "The Nature Conservancy is a great partner for The Home Depot in this effort," Jarvis said.

The gift from The Home Depot augments vital funding provided since 2001 by the U.S. Agency for International Development for the conservation of orangutan habitat, combating illegal logging and promoting sustainable forest practices through work with the timber industry, local governments and indigenous Dayak villagers. Additional funding, provided by the U.S. Fish and Wildlife Service and private donors, is paying for further orangutan surveys and other key habitat conservation efforts.

The Conservancy's orangutan survey plan was developed by Harvard University primate expert Andrew J. Marshall, utilizing established scientific survey methods. Because spotting the rare and elusive orangutan is difficult and takes a prohibitively long time, Conservancy consultants recruited and trained indigenous Dayaks to identify and count orangutan nests, a widely accepted method for assessing the size of orangutan populations. Using the survey results, Marshall then calculated the size of the orangutan population.

"Given that conservation funds are always limited and that political support and logistical constraints vary in different places, it is crucial that financial resources be focused on areas where the chances of protecting viable orangutan populations are greatest," Marshall said. "The orangutan habitat area in East Kalimantan is one of those places."

Other orangutan experts echoed the Conservancy's excitement over this find.

"The discovery of a large, biologically viable, previously unsurveyed orangutan population in East Kalimantan is very significant," said Dr. Birute Mary Galdikas, president of Orangutan Foundation International. "This find extends the orangutan's known range and gives us hope that we can save orangutan populations from extinction in the wild."

## The Nature Conservancy Uses Computerized Bar Codes to Stop Illegal Logging in Indonesia

from The Nature Conservancy

In a move to stop illegal logging rampant in Indonesia, The Nature Conservancy . . . launched an innovative project using computerized bar codes to track timber coming from the country's forests.

Similar to the ones you see in grocery stores, the bar codes will be placed on trees cut legally by two Indonesian companies who have volunteered to work with the Conservancy on the project.

The bar codes will allow the timber to be tracked from stump to store so consumers can choose products they know are legal. This includes compliance with environmental laws so as to help reduce threats to forest habitats.

"Illegal logging is decimating Indonesia's forests," said Steve McCormick, president of The Nature Conservancy. "Indonesia's illegal logging hurts the global environment, threatens endangered animals such as orangutans and rhinos, and harms local communities that depend upon healthy forests for their livelihoods."

He added: "This bar coding technology will allow consumers to support companies who do the right thing, and put market pressure on other companies to stop illegal logging."

The three-month pilot project will cost $400,000 and is being funded by the British Department for International Development, the US Agency for International Development, and The Home Depot—the world's largest retailer of wood.

"As the world's largest retailer of wood, The Home Depot wants to promote legal and sustainable forestry," said Ron Jarvis, The Home Depot's merchandising vice president for lumber and building materials. "Consumers care about illegal logging and, through this program they can use their wallets to help stop it."

Under this project, bar code tags will be placed on legally-harvested timber, giving it "digital fingerprints." The tags carry a unique number which corresponds to database information about a log's size, species and origin, making it almost impossible to swap tags. The tags follow the timber through processing and manufacture.

The bar codes allow external auditors to scan thousands of logs quickly, to compare whole barges or stockyards with a readily updated database.

Some 75 percent of logging in Indonesia is illegal, and many countries and companies—including Britain and The Home Depot—have announced they will boycott illegally-logged timber.

Indonesian logging companies are beginning to worry that these boycotts will cost them millions of dollars, so they have begun to work with environmental groups like The Nature Conservancy to ensure that their timber is harvested legally—and that consumers know about it.

"If buyers can tell the difference between what's legal and what's illegal, we can use market forces to send the right signal to logging companies" said Nigel Sizer, director of The Nature Conservancy's forest conservation programs in Asia and the Pacific, who also conceived the system. "We had to develop a tool to allow buyers to make that distinction.

The two Indonesian companies that are now participating in the project are PT Sumalindo Lestari Jaya and PT Daisy. The Nature Conservancy is talking with other Indonesian logging companies and is hopeful that five to 10 more will join the project in the next year.

"A growing portion of the market is demanding this or something similar and it will soon become a basic part of business for companies in Indonesia and elsewhere where illegal logging is prevalent," Sizer predicted.

# Do Programs Developed by Conservation Organizations to Protect Habitat and Endangered Animals Work?

Outsiders telling native people that they can't log their forests or that they can't hunt local animals to sell at great profit has not proved to be a successful way to protect the land and its creatures. But people can be taught to use their resources in a way that sustains both the people and the resources. This means logging in a way that maintains forests and hunting in a way that allows animal populations to survive. Conservation organizations—like the World Wildlife Fund (WWF) and the Wildlife Conservation Society (WCS)—have developed programs to give native peoples economic reasons to protect the resources. The following three articles give examples.

The first article provides an update on the Hyacinth Macaw Project in the rain forests of Brazil. In the 1980s, more than 10,000 hyacinth macaws were stolen and sold into the multibillion-dollar global animal smuggling business. In addition, the birds' habitat was, and still is, being destroyed. In 1990, ornithologist Neiva Guedes received a grant from the WWF to protect and restore the endangered population of the hyacinth macaws, which are the world's largest parrots. Working with landowners, communities, and tourists, Guedes has seen the population of birds double. But in other forests, they still remained threatened.

The second article discusses the conservation program of the New York–based Wildlife Conservation Society (WCS) to restore a population of Cambodian river turtles. The project works with local people, many of them former Khmer Rouge soldiers, to guard new nests and to teach villagers about the importance of protecting the turtles. Using the approach of working with local people has increased the number of these turtle hatchlings swimming to the sea.

The third article shows how a Cameroon pygmy community in Central Africa has benefited from a WWF hunting program. To fight the growing problem of villagers killing monkeys and other animals to sell, the WWF worked with a local timber company to set up responsible hunting programs for foreign trophy hunters. With the huge profits from this, the villagers are able both to make a substantial profit and to sustain their resources.

Many of these programs involve the partnering of conservation organizations, businesses, local governments, international agencies, and native people. The goal is to make it economically attractive to protect important habitats around the world. And all over the world, the number of success stories continues to grow. Whether endangered species will ultimately survive remains a question of whether additional programs can be developed in time.

—The Editor

## The Hyacinth Macaw Makes a Comeback

by Harold Palo

A hot June day in the Pantanal, Brazil. It's the dry season. Although this is the world's largest tropical wetland, the grass is now yellow, and there's hardly any water to be seen. The few ponds that do remain after the wet season floods are full of caimans. A group of nandus have found shade under some trees on the rolling fields. I'm on my way to visit the Hyacinth Macaw Project at the Refúgio Ecológico Caiman, some 250 kilometers [155 miles] west of the city of Campo Grande.

Entering the project office, I see a poster showing all the parrots and macaws of Brazil. Four completely blue macaws catch my eye. Curious, I point to the first one.

"We call this Arara azul pequena, *Anodorhyncus glaucus* in Latin," says Cézar Corrêa, the project's Research Assistant. "It became extinct in 1950."

I point to the second blue macaw.

"Ararinha azul, *Cyanopsitta spixii.* It became extinct in the wild in 2000. Around 60 birds remain in captivity."

I point to the third.

"Arara azul de Lear, *Anodorhynchus laeri*, Lear's macaw in English. Around 450 still live in the wild and some in captivity."

Three of these four beautiful blue birds gone or almost gone. I am shocked.

I point at the last one, the hyacinth macaw. It's the largest of them all, measuring 1m from beak to the tip of the tail, with a wingspan of 1.5 m [1.64 yards]. I've traveled thousands of kilometres to see it, *Anodorhynchus hyacinthinus*, or Arara azul as it's called in Brazil.

"We estimate 6,500 hyacinth macaws remain in the wild, of which around 5,000 live in the Pantanal." I sigh with relief.

But not long ago the hyacinth macaw, the world's largest parrot, was also in great danger. In the 1980s an estimated 10,000 specimens were illegally captured and sold as pets, mainly on the international black market. A single bird could bring in $12,000 [in U.S. dollars].

On top of this, the species' natural habitat was being destroyed by deforestation, burning, and planting of pasture for cattle. Local Indians used fire weapons to kill the macaws, taking their feathers to make souvenirs for tourists. By the end of the 1980s, only 2,500–3,000 remained in the wild.

Then, in the 1990s, the Brazilian Pantanal population of hyacinth macaws recovered. In one 4,000 km$^2$ area, the number doubled within 10 years—from 1,500 in 1990 to 3,000 in 2000. How did this come about?

At the end of the 1980s, 27-year-old biology student Neiva Guedes was on tour in the Pantanal. Watching a flock of large, deep-blue macaws flying by, her professor said: "These hyacinth macaws are likely to become extinct during our lifetime."

Neiva was struck. At that very moment the course of her life changed forever. Determined not to let this extinction happen, she started the Hyacinth Macaw Project.

Cézar, a former car mechanic, is not a man you will run down easily. Strong and supple, with a short goatee and dark eyes under a hat, he seemed pretty tough in the project office. But I soon discover he's a man full of love.

Cézar drives me across the rugged terrain of the Refúgio Ecológico Caiman. It's a place of permanent and temporary lakes and ponds, with rolling fields of grass and shrub broken by corridors and islands of forest.

Wooden fences indicate this is cattle country. It's also the land of jaguar, ocelot, caiman, tapir, and giant anteater. More than 600 bird species live here too—ibis, storks, toucans, and, of course, macaws.

There they are, three hyacinth macaws! Two adults and a young. "A family," Cézar explains. "The young stay with their parents for around 18 months."

I train my binoculars on them. Blue, so very blue! Golden eye rings and cheeks. There they go, flying away with power and grace, their long tail behind them.

In the Pantanal, hyacinth macaws—highly social and faithful birds that mate for life—prefer to make their nest in the *manduvi* tree, whose soft trunk is easily hollowed out by a macaw beak. In the process of enlarging natural cavities, the birds also create a lining of small woodchips and sawdust for the eggs to rest on.

We arrive at the first nest, about 8 m [8.75 yards] above the ground. A small rope hangs down from a branch higher up. This nest has been monitored before.

After attaching a longer rope to the one on the tree, Cézar snaps on a harness seat. Stretching my neck, I watch him quickly haul himself up. He puts a hand into the cavity and feels around.

"The birds are preparing their nest," says Cézar after he's come back down. "I felt the wood chips." If the cavity had been too deep for the future young to get out, Cézar would have added extra woodchips to raise the floor of the nest.

Two black vultures sit on guard next to the second nest we visit. A hyacinth macaw appears in the opening of the cavity. Head slightly tilted, it calmly watches the predators. This macaw has no invasion to fear. Cézar has fixed boards of wood around the nest opening, making it too small for the vultures—and other predators like hawks and large owls—to get in.

The third nest we visit is not a cavity at all, but rather a wooden box placed high up in a tree. Cézar constructs these artificial nests because there are too few natural nesting sites.

These interventions were all developed by Neiva Guedes. With support from the University for the Development of the State and Region of the Pantanal, she created the Hyacinth Macaw Project in 1990. She then taught herself how to climb trees and began monitoring the macaws' nests and chicks.

Neiva found that the survival rate of hyacinth macaw chicks is generally low. Breeding pairs lay two eggs on average, but usually only one survives. The eggs and chicks are often taken by predators, and also, the second chick will not survive if it hatches more than four days after the first.

Neiva's ideas to increase the breeding success of the hyacinth macaw have been very successful. Artificial nests and the use of boards to keep predators out of natural nests have contributed considerably to the species' recovery in the project area.

Chick management has also been effective. In nests that have a history of unsuccessful breeding, Cézar replaces the macaw eggs with chicken eggs. The macaw eggs are incubated in the field laboratory. After hatching, the chicks are fed for a few days and then reintroduced to the original nests or to another nest with chicks of the same age.

Despite the success of these measures, some disapprove of interfering like this.

"Artificial nest boxes are a short- and medium-term measure," counters Neiva. "We use them because one of the biggest problems for hyacinth macaws is a lack of natural nests. We also work with the ranchers for long-term solutions, like preserving native trees such as the *manduvi*."

"We only manage chicks in nests where there would be no chick survival without our interference," she continues. "And before we do this, we first limit ourselves to managing the nest, for example by making the opening smaller to protect against predators."

Neiva is the hyacinth macaw's equivalent of Jane Goodall. Besides being a gifted researcher, she's also a successful "warrior for the hyacinth", as one cattle rancher puts it. In between collecting data over the past decade, Neiva has tirelessly paid visits to the cattle ranches in the region, raising awareness of the birds and what they need to survive.

Expanding cattle ranches—whose pastures replace the macaw's natural habitat—are the major threat to the species today in the Pantanal. The highly specialized birds have a high-energy diet, based on the nuts of two palm tree species, *acuri* and *bocaiúva.* Safeguarding these two feeding trees plus the macaw's favourite nesting tree, the *manduvi*, is a priority.

Thanks to Neiva, ranchers are now beginning to be proud to have a macaw nest on their property. The awareness raising has also diminished the threat of the illegal trade in hyacinth macaws in the project area.

"The Hyacinth Macaw Project has brought hope to this species in the Brazilian Pantanal. The results are outstanding," says Bernadete Lange from the Brazilian office of the global conservation organization WWF, which has supported the project since 1999.

"But unfortunately, the species is not doing so well in other areas of Brazil, such as the Cerrado savannahs and the eastern Amazon," says Bernadette. "Seventeen years ago there were 1,500 individuals in these areas, but today only 1,000 remain. This is due to conversion of their forest habitat to pasture and soy crops. The Brazilian government must treat the hyacinth macaw as an endangered species, and work to protect its natural habitat."

One incentive for the government and local landowners to protect wildlife and natural habitats could be ecotourism. The Pantanal is one of the best places in the world to watch birds and animals. Neiva and her team are thinking of organizing field trips for ecotourists to experience the hyacinth project.

The money from such trips is desperately needed. The staff of the Hyacinth Macaw Project lack people and resources to adequately monitor macaw nests. More money is also needed to properly preserve macaw habitat.

Despite the difficulties, Neiva remains full of passion.

The purpose of my work, which means my life, is to preserve the hyacinth macaw in the wild," she says. "I don't care about having 100, 200, or 300 birds in captivity, 50 or 100 years from now. I care about a sustainable population of hyacinth macaws flying free in Brazil."

## A Royal Return

by Karen Coates

The humid air hangs like smoke as we putter up the Kaong River in rural southern Cambodia. A parakeet swoops low and butterflies cluster on the shore, their erect wings clasped tightly. The afternoon sun bathes everything in hot white light.

Soon we arrive at a small beach that looks like any other along the Kaong, except for four bamboo cages planted in the sand. They surround the eggs of an endangered giant terrapin, Batagur baska, which conservationists have transported from a spot upriver where high tides would have washed them away. "They cannot hatch in the water," explains Heng Sovannara, a Cambodian biologist from Phnom Penh.

Moving these nests is a slow, meticulous process—each egg must be recorded and placed in the new nest precisely as it was laid. More than 100 days old, the eggs are now ripe for hatching. After three men dismantle a cage top, Sovannara, who works for the New York–based Wildlife Conservation Society (WCS), clambers inside. He paws through sand, flinging golden grains, digging six inches down. "Yes!" he shouts. "Hatch already!"

At the bottom of the nest is a miniature Batagur baska, legs flailing—boom, on its back, then upright again, trying to escape the hole. The reptile is less than three inches long and has a yellowish shell. Sovannara holds the tiny turtle as it squirms in air, and the audience breathes a collective, joyous "Ahhhhh . . ."

A hatching *Batagur baska* hasn't always been such a monumental event. Generations ago, legions of these giant terrapins swam Cambodia's rivers, swamps and estuaries. Considered property of the country's royal family, the king hired guards to protect the animals. "A long time ago, the people collected eggs to give to the king," Sovannara says. "Only the king could eat these eggs"—which is why *Batagur baska* is called the royal turtle in Cambodia.

But despite the king's decree, royal turtles were hunted over the years, their eggs collected and their habitat destroyed. "They killed all," Sovannara says of the local villagers. Until 2001, when a small colony was discovered along the Sre Ambel River system in former Khmer Rouge territory, scientists believed *Batagur baska* had been extinct in Cambodia for the past century.

Now, thanks to Sovannara and other Cambodians who participate in a WCS project—many of them former Khmer Rouge soldiers—the royal turtle may have a second chance. Launched in 2002, the project began with a survey of terrapins in the area, followed by the creation of a year-round program to guard new nests, deter poachers and teach villagers about the animal's plight. "The amazing commitment both the local village rangers and the government have shown at protecting Batagurs and their nests makes me optimistic," says WCS Cambodia program manager Colin Poole.

One of the largest in the Emydidae family of turtles, *Batagur baska* can grow to more than two feet in length. Its olive gray carapace blends in well with the murky waters it inhabits in Southeast Asia. The turtles, which may live to be 100, reach sexual maturity at about 25 years old. Mating just before or early in the dry season, females lay between 13 and 34 eggs three months later and may nest up to three times a year. The turtles compact their nests by thumping their bodies on sand—making a noise that gives the species the name tuntung in Malay.

*Batagur baska* once inhabited rivers and swamps from India east through Vietnam and south through Indonesia. But today only small populations remain in Malaysia, India and Bangladesh, as well as the Sre Ambel River region of Cambodia. IUCN—The World Conservation Union lists the reptile as a critically endangered species.

Batagur populations were decimated by a combination of threats, including water monitor lizards, nets and explosives used by fishermen, logging and weak law enforcement. Throughout its range, the turtle's habitat dwindled as forests were cut and boats plied rivers where it once thrived. The terrapins also taste good. "Delicious for the villagers," Sovannara says. Eggs are a delicacy, and the animals' meat—like that of other endangered turtles—is still a prized commodity, not only in Cambodia but throughout Asia.

Heading up the Sre Ambel River the following day, we stop at the home of a ranger named Osmang. In his small thatch house on stilts, a Panasonic radio hangs from a post, burbling old American love songs. We sit on a slat floor with Osmang, his wife and a friend. Their possessions comprise the Panasonic—new this year, paid for by the turtle work—cooking utensils, a fish net and numerous cassava, chili and papaya plants surrounding the home. The couple has five children, three girls and two boys, ages 5 to 19.

Osmang, 41, says he ended his work as a Khmer Rouge soldier before Cambodia's 1993 United Nations–sponsored elections. When I ask his thoughts on his previous life, his wife, Kaow, abruptly answers for him: "Didn't like it! Didn't like it!"

Osmang agrees: "This job is better," he says. "I didn't like to work with the Khmer Rouge soldiers, but there was no way to choose because the government was the Khmer Rouge and the Khmer Rouge were the government." He, like many others, was forced to fight. Osmang also admits that he previously ate turtles. "Every year I hunted to bring food to my children," he says. But now he works to save the reptiles.

Though its primary purpose is to protect and restore *Batagur baska*, the turtle project also nurtures local people by hiring former Khmer Rouge soldiers—the safest, most lucrative jobs they've ever had. The conservation team includes seven former soldiers, hired as rangers or turtle collectors, plus a speed boat driver and team leader from the Fisheries Department in the town of Sre Ambel. WCS pays each team member $40 a month. All participants teach local villagers and fishermen how to recognize royal turtles—and why it is important to leave the animals and their eggs alone.

Their work seems to be paying off. According to Sovannara, royal turtles no longer appear in the markets around Sre Ambel. Yet in a country with an average per capita income of $260 a year and rampant poverty, people still sell and eat *Batagur baska* meat and eggs elsewhere. Sovannara says the reptile's meat sells in the market for 2,000 riel, about 50 cents, per kilogram; the eggs sell for 1,000 riel, or 8 cents, apiece.

Some of *Batagur baska*'s biggest threats lie outside Cambodia. In China, food markets sell thousands of endangered Asian turtles every year, according to a report by James Barzyk for the London-based Tortoise Trust. In 1996 alone, 7.7 million pounds of turtles were eaten in Hong Kong. In December 2001, 10,000 live turtles were seized from a ship stopped offshore. According to officials, the animals had been caught in the wild and many were injured during shipment. The lot contained 11 species, including *Batagur baska*, worth just over $400,000.

In the report, Barzyk describes a "three prong attack" against endangered Asian turtles: collecting adults, harvesting eggs and destroying habitat. He blames these threats for the royal turtle's 90 percent decline over the past century in Malaysia alone. In Thailand, the species is probably extinct in the wild, according to Kalyar, a biologist at Chulalongkorn University in Bangkok.

Yet here along Cambodia's southern rivers, there is hope. In 2003, seven protected nests yielded 59 squirming hatchlings, and biologists set free 31 hatchlings the previous year.

That night at Osmang's house, we sleep beneath mosquito nets, with the moon rising as a misty orb over far-off trees. The sound of terrapin claws against aluminum pots serenades us like soft rain. Rangers placed the tiny turtles in the pots earlier in the day. The animals tumble against each other, scratching, clawing, pushing in an instinctive search for water. They will do this all night and all morning until their release in higher waters on the Sre Ambel.

Early next morning on the river, the air is sweet with jasmine as we motor slowly, passing an otter in our path. We disembark at a grassy path near a highway and head to Sre Ambel town. There, Sovannara dangles each hatchling in a plastic bag clipped to a hand-held scale. With a measuring tape, he notes length and width. Then we're off to set the turtles free.

Up the Sre Ambel, we stop in Kring Chek village. Sovannara's first task, as always, is to educate the crowd ogling the terrapins. Villagers say they have seen soft-shelled turtles before, but nothing like these.

Then, we hike through a field to a mucky spot at the water's edge. Everyone wants a part. We're covered in mud, up to our ankles, calves, thighs. People dig into the pot, each grabbing a turtle, and—plop—into the water they go. They sink, pop up, then spin around before entering the current. Children squeal with delight.

Some turtles return toward land, as though confused, but villagers redirect them. Tiny heads bob in straight lines across the water. "I am very happy the baby can swim across the river," Sovannara says. "I request to the villagers, 'Help us protect this species.' I think in the future the population will increase."

Back in the boat, pulling away from the beach at Kring Chek, we see a man up to his chest in murky water, fishing net in hand. A woman riding with us hollers at him in Khmer. Words burn through her mouth in amazing pitch, volume and speed for several minutes, while the man silently nods his head: "Yes, yes, yes."

"What did she say?" we ask Sovannara in English.

"She said: 'Don't catch.' "

## Hunting for Conservation

from the World Wildlife Fund and Wildlife Conservation Society

On the edge of the rainforest in southeast Cameroon, Baka pygmies from the village of Lantjoue are having a party. Everyone is dressed in their finery and the drums are beating. A feast of yams, plantains, freshly caught fish, and a big pot of antelope stew is spread out in the village school room.

The reason for the party? The arrival of new desks and equipment for the school, paid for by a community hunting project that the villagers set up with help from global conservation organization WWF.

"It may seem surprising to find a conservation organization supporting hunting," says WWF's Leonard Usongo, the local manager of the project. "But commercial hunting for bushmeat has become such a problem here that we had to try something new to control it. One approach is regulated hunting."

Lantjoue is typical of the small communities on the fringes of Cameroon's rainforest. The Baka and Bantu people have lived here for generations, eking out a living by growing crops, working in the logging concessions, and hunting and gathering in the forest.

"The people here have always hunted for their own needs," says Usongo. "But in the last couple of decades new roads have been opened, mostly for logging, and there are lots more trucks heading for the cities. Local hunters can sell bushmeat to passing truck drivers for more money than they could ever have dreamed of a few years ago. This has fuelled a huge increase in hunting, including some animals that are endangered—like gorillas."

The truck drivers sell the meat in the markets of Yaoundé and Douala, where it commands a high price. The trade is so lucrative that it has attracted people from other parts of the country, who now poach animals in the forests.

"We tried working with Cameroon's Ministry of Environment and Forests to stop the trade," says Usongo, "but there are too many trucks and too many roads—it's impossible."

The new approach is to help local people manage hunting for themselves. Instead of government-imposed rules and penalties aimed to discourage hunting for the bushmeat trade, the villagers of Lantjoue can instead regulate their own hunting quotas in a defined village hunting zone.

One incentive to keep wildlife abundant is foreign trophy hunters.

Among many other species, the forest around Lantjoue is home to the elusive bongo antelope (*Tragelaphus euryceros*). This magnificent tan-and-white-striped animal with slender spiralling horns only lives in a few places in Africa. Trophy hunters are prepared to pay large sums of money for one.

These rich foreigners want to be sure that they will find a bongo during a fairly brief visit. If the villagers can guarantee this, then the trophy hunters will come to their forests.

Under the project set up by WWF, the villagers must limit their own hunting and ensure that lots of bongos can be found in their forest. The trophy hunters pay a large license fee, part of which is returned to the villagers to pay for improvements such as the equipment for the school.

Diopim Akanda, the village chief, is happy.

"As long as we can keep outside poachers away, we can find enough animals for our food and still attract the foreign hunters, who pay us more than we could get selling bushmeat to passing truck drivers."

A small group of Baka pygmies have set up a camp next to the village, and act as guides for the trophy hunters.

"The pygmies have an astonishing knowledge of wildlife," says Usongo. "It's fascinating to spend a day in the forest with them. You see things that you would never see on your own, and they understand the habits of the animals amazingly well. There are gorillas, chimpanzees, and a wealth of other species to be seen. We hope that in the future, ordinary tourists will come to shoot with their cameras rather than with guns."

Adjacent to the village hunting zone is a large logging concession run by a Belgian family. Manager Jules Decolvenaere has also joined forces with WWF.

"We are keen to get our timber certified under the Forest Stewardship Council (FSC)," he says. "We think that we already meet most of their standards for environmentally and socially responsible forest management. We also try to protect wildlife but it is very difficult, and conservation groups will criticize us if poachers come into our concessions."

Decolvenaere welcomes the new village hunting zone and supports the WWF initiative.

"If we can demonstrate that hunting is under control in the area then we will be taking a big step towards certification. This will give us access to better markets in some European countries."

It's also a matter of professional pride for Decolvenaere.

"My family has been working these forests for over 30 years," he says. "We are keen to demonstrate that our industry can be good for the forest and good for the local people."

To help the efforts to restrict hunting, the logging trucks returning from the cities now bring frozen meat back to the concessions. Decolvenaere wants to be sure his workers have access to meat without resorting to poaching.

"We pay our staff well and we want them to share our goal of being a responsible environmental company—so we practice good logging and we try to protect the wildlife."

Leonard Usongo is enthusiastic about the new developments.

"We used to put all our efforts into national parks but it was difficult to get much local support," he says. "This area is too remote for most tourists so the parks don't do much for the local economy. Now we are trying to conserve the broader landscape.

"The national parks still exist of course. But now we also work with concessionaires to improve the management of logging operations and with local people to ensure they can get jobs and also continue to harvest the things they need from the forest."

Jill Bowling, who manages WWF's global Forest Programme, believes the work in southeast Cameroon has potential in other parts of the world.

"If we want our conservation programmes to be sustainable in the long-term then they have to make sense to local people," she says. "Just setting aside vast areas of forest and closing them to people cannot work."

WWF's approach now emphasises a balance between protecting, managing, and restoring forests—which makes a lot more sense to local partners in poor countries than just protection alone.

Diopimb Akanda agrees.

"All our traditions and culture are linked to the forest," he says. "So we care about the forest—but we also want education, jobs, and health clinics. And if the local economy doesn't thrive then our children will move to the cities and only the old people will stay here."

"Thanks to this project, we can find work in the concessions, we can guide the trophy hunters, and we can still hunt for our own needs," he adds. "We hope in the future that more tourists will come and that we will be able to share with them our knowledge of the forests and our culture."

# BIBLIOGRAPHY

Coates, Karen. "A Royal Return." *National Wildlife Federation.* April–May 2003.

Fascione, Nina. "America's Wolves Threatened Again." *Defenders of Wildlife.* Spring 2003. Available online at *http://www.defenders.org/defendersmag/issues/spring03/wolves.html.*

The Nature Conservancy. *The Nature Conservancy Finds Population of Wild Orangutans.* 2002. Available online at *http://nature.org/pressroom/press/press858.html.*

———. *The Nature Conservancy Uses Computerized Bar Codes to Stop Illegal Logging in Indonesia.* 2004. Available online at *http://nature.org/pressroom/press/press1520.html.*

———. *Summary of Orangutan Surveys Conducted in Berau District, East Kalimantan.* Available online at *http://nature.org/magazine/winter2002/orangutans/.*

Nickens, T. Edward. "The Art of Helping Wildlife." *National Wildlife Federation.* February–March 2003. Available online at *http://www.nwf.org/nationalwildlife/dspPlainText.cfm?articleId=732.*

Pew Oceans Commission. *Marine Reserves: A Tool for Ecosystem Management and Conservation.* 2002. Available online at *http://www.pewoceans.org/oceanfacts/2003/01/13/fact_31395.asp.*

Semlitsch, Raymond D., ed. *Amphibian Conservation.* Washington, D.C.: Smithsonian Books, 2003, pp. 5–7.

U.S. Fish and Wildlife Service. *Endangered Species Bulletin No. 4.* Vol. XXVII. July–December 2003. Available online at *http://endangered.fws.gov/esb/2003/07-12/toc.html.*

Wilson, E. O. "Biodiversity." *Audubon.* December 1999. Available online at *http://magazine.audubon.org/biodiversity.html.*

World Wildlife Fund. *CITES: Ensuring That Species Are Not Threatened by International Trade.* Available online at *http://www.panda.org/about_wwf/what_we_do/species/our_solutions/wildlife_trade/cites.cfm.*

———. *Conserving Nature: Partnering With People, World Wildlife Fund's Global Work on Protected Area Networks.* 2003. Available online at *http://www.panda.org/downloads/wpcbrochure.pdf.*

———. *Flagship Species Factsheets.* Available online at *http://www.panda.org.*

———. *The History of Whaling and the International Whaling Commission.* 2004. Available online at *http://www.panda.org/news_facts/factsheets/species/publication.cfm?uNewsID=13796&uLangId=.*

———. *Hunting for Conservation.* July 2004. Available online at *http://www.panda.org/news_facts/newsroom/features/news.cfm?uNewsID-14073.*

———. *The Hyacinth Macaw Makes a Comeback.* April 2004. Available online at *http://www.panda.org/about_wwf/what_we_do/freshwater/stories/news.cfm?uNewsID=12501.*

———. *What Has It Achieved So Far?* Available online at *http://www.panda.org/about_wwf/what_we_do/species/our_solutions/wildlife_trade/cites_defined.cfm.*

Youth, Howard. *Winged Messengers: The Decline of Birds.* World Watch Institute Paper #165. 2003.

## FURTHER READING

Hunter, Malcolm L. *Wildlife, Forests, and Forestry: Principles of Managing Forest for Biological Diversity*. Englewood Cliffs, NJ: Prentice-Hall, 1990.

Quammen, David. *Song of the Dodo*. New York: Scribner, 1996.

Semlitsch, Raymond D., ed. *Amphibian Conservation*. Washington, D.C.: Smithsonian Books, 2003.

Wilson, E. O. *Biophilia*. Cambridge, MA: Harvard University Press, 1986.

———. *Diversity of Life*. Cambridge, MA: Harvard University Press, 1992.

———. *Future of Life*. New York: Random House, 2002.

## WEBSITES

**Defenders of Wildlife**
http://www.defenders.org/

**National Wildlife Federation**
http://www.nwf.org/

**The Nature Conservancy**
http://nature.org/

**Pew Oceans Commission**
http://www.pewoceans.org/

**U.S. Fish and Wildlife Service**
http://endangered.fws.gov/

**World Wildlife Fund**
http://www.panda.org/

# INDEX


Africa
great apes in, 55, 88. *See also* chimpanzees; gorillas; orangutans
African elephant, 51–53
Agassiz National Wildlife Refuge, 83
Air pollution, xi–xii
Air quality, xviii
Akanda, Diopimb, 111
'Akiapol'au, 84
Alabama, 36
Aleutian Canada goose, 35
Algonquin Park (Canada), 47
Alliance of Religious and Conservation (ARC), 70
Amazon rain forest, 5–6
American bullfrogs, 11
American Lung Association, xii
American Museum of Natural History, 3
*Amphibian Conservation* (Semlitsch), 9, 10–12
Amphibians
conservation of, 12
decline in, 9
importance of, 9
reasons for loss of, 10–12
*Ants, The* (Wilson), 3
Apes, 88. *See also* chimpanzees; gorillas; orangutans
Aquatic species, 36–37
Aransas National Wildlife Refuge (Texas), 84
Arctic National Wildlife Refuge, xxiv
Arkansas, 34–35
Asian birds, 18
Attwater's greater prairie-chicken, 38–39
Attwater's Prairie-Chicken National Wildlife Refuge, 38
Audubon Society, xi, xxiv
Australia, 18

Baja California, 57
Baka pygmies, 108, 109, 110
Bald eagle, xiv–xv, 33
Bamboo "die back," 50–51
Bangladesh, 105
Bar codes, 94–96
Barr, Brian, 45–46
Barzyk, James, 106
Batagur baska, 104–107
Beef, imported, 7
Bekko shell, 58
Bengal tigers, 56
Berau District, East Kalimantan, 89–91, 92, 93
Bighorn sheep, 82
Biodiversity
American understanding of, 8
defined, xvi
environmental activism to protect, 6–7
E. O. Wilson and, 3
future of, 4–5, 8
impact of individual actions on, 7–8
levels of, 2
loss of, impact of, 4, 5
preserving, 2
in Yellowstone National Park, wolves and, 44–46
BirdLife International, 15, 18
Birds
Attwater's greater prairie-chicken, 38–39
benefits and contributions of, 13, 16
decline in, xi, 14–15
discovery of new species, 18–19
extinction of, 13
human-related factors threatening, 17
hyacinth macaw, 98–103
inspiration from, 15–16
regional estimates of decline in, 18
threats to, 13
unsurveyed, 17–18
Blue whales, 25
Bongo antelope, 109
Borneo, 89, 91–92
Boulder dart, 36
Bowling, Jill, 110, 111
Brackett's Landing Shoreline Sanctuary Conservation Area (Washington), 75–76
Brazil, 98–103
British Department for International Development, 95
British Petroleum (BP), xiii
Browner, Carol, xii
Bureau of International Statistics, 24
Bushmeat, 88
gorillas and, 54
impact on chimp population, 53
indigenous peoples and, 108–111
Business practices, xiii, 68–69

Caecillians. *See* amphibians
California
Carson wandering skipper in, 37
marine protected areas in, 75
marine reserves in, 76
Cambodia, 103–107

# INDEX


Cameroon, 108–111
Canada, 47, 84
Cane toads, 11
Cap-and-trade systems, xii–xiii
Carolina Sandhills refuge, 84
Carson, Rachel, xix, xxiii, 13
Carson wandering skipper, 37
Carter, Jimmy, xxiv
Cattle ranches, 102
Central chimpanzees, 55
Chapin, Ted, 46
Chemical contamination, amphibian decline and, 11
Chimpanzees, 53, 55, 88
China, xv, 49, 106
Chinese medicine, 56
CITES (Convention on International Trade in Endangered Species), xv, xxvi
  components of, 31
  conferences on, 31–32
  history of, 30–31
  individual actions and, 7
  ivory trade and, 52
  marine turtles and, 58
  World Wildlife Fund and, 68
Clean Air Act (1970), xi–xii, xviii
Clean Air Rules of 2004, xviii
Clean Water Act (1972), xxi
Climate change, xii, xviii–xix
  amphibian decline and, 10
  marine turtle population and, 58–59
  World Wildlife Fund strategies on, 67
Clinton, Bill, 88
Coal, burning of, xvii
Coates, Karen, 103
Coffee, shade-grown, xiv, 7–8
Colorado, 46
Columbia Basin pygmy rabbit, 38
Commercial trade. *See* trade
Conference of the Parties (CoP), 31
Congo, 54–55
Conservation, xix–xx. *See also* protected areas; individual actions
  of amphibians, 12
  bushmeat hunting and, 108–111
  elephant, 52–53
  of hyacinth macaws, 98–103
  international efforts for, xii–xiii
  public recognition for efforts at, 70
  of royal turtles, 103–108
  U.S. history of, xxiii–xxiv
  whale, 27–29
  World Wildlife Fund strategies for, 64–65
Conservation International, xxiv, 3
Conservation organizations. *See also* individual names of organizations
  diversity of, xxiv
  preserving biodiversity and, 6–7
*Conserving Nature: Partnering With People* (World Wildlife Fund), 62, 63–71
Convention for the Regulation of Whaling, 24
Convention on Biological Diversity, 67–68
Convention on International Trade in Endangered Species of Wild Fauna and Flora. *See* CITES
Convention on Migratory Species Convention, 68
Coral reefs, 59
Corrêa, Cézar, 98, 99–101
Costa Rica, 7
Cross River gorilla, 88
Cuban treefrogs, 11
Cuyahoga River, xxiii

Dayak villagers, 93
Debt-for-nature swap programs, xxvi
Decolvenaere, Jules, 109, 110
Deforestation. *See* logging
De Hoop reserve (marine), 76
Desert bighorns, 82
Desert National Wildlife Refuge, 82
Dippers, 16
Disease
  amphibian decline and, 11
  great ape population and, 54–55
  in marine turtles, 59–60
*Diversity of Life* (Wilson), xvi
Douglas County (Washington), 38
Ducks, 81–82
Dutton, Donna, 80, 81

# INDEX


Earthwatch Institute, 80
Ebola crisis, 54–55
Ecoregions, 64
Edmonds Underwater Park, 75–76
Eldredge, Niles, xi, 36
Elephants, 51–53
Elk, 44
Elk River, 36
Ellis, Gerry, xv
El Niño–Southern oscillation, 10
Ely, Minnesota, 47
Emerson, Ralph Waldo, xxiii
Endangered species. *See also* amphibians; birds; turtles; whales; wolves
  aquatic species, 36–37
  Carson wandering skipper, 37
  gopher frog, 38
  in Hawaii, 37
  listed on Endangered Species Act, 34–39
  Magazine Mountain shagreen, 34–35
  pallid sturgeon, 37–38
  prairie bush clover, 34
  prairie chicken, 38–39
  pygmy rabbit, 38
  recovery of, 36
  riparian brush rabbit, 35
  whale, 22
Endangered Species Act (ESA) (1973), 33
  purpose of, 34
  species listed on, 34–39
  wolves and, 40, 41–42
*Endangered Species Bulletin* No. 4, 34–39
Environmental Defense, xxiv
Environmental policy, xx–xxi
Environmental Protection Agency (EPA), xii, xviii, xix, xxiv
European bird species, 18
Everglades, the, x
Extinction, x–xi
  bird, 13, 15, 17
  Endangered Species Act and, 33, 34–39
  future of, 4–5
  of macaw birds, 99
  wave of, 17, 36
  whale, 22

Fascione, Nina, 41
Fish
  amphibian decline and, 11–12
  pallid sturgeon, 37–38
Flagship species, 48, 66–67
Flooding, 82–83
Florida Keys National Marine Sanctuary, 76
Food markets
  Asian turtles and, 106
  bushmeat and, 53, 54, 108–111
  individual actions and, 7
Ford, William, xiii
Forests
  protection of orangutans and, 93
  restoring, on refuges, 83–84
  World Wildlife Fund protection of, 65–66
Forest Stewardship Council (FSC), 110
Frogs, endangered, 38. *See also* amphibians

Galdikas, Birute Mary, 94
Game fish, 11–12
Geese, 81–82
Germany, 24
Giant panda, xv, 48, 49–51
Gibbs, Lois Marie, xxiv
Gift to the Earth scheme, 70
Global 200 Ecoregions, 6, 65
Global warming. *See* climate change
Gopher tortoise, 38
Gorillas, 54–55
Government practices, xxiii, 68
Gray wolf, 33, 35. *See also* wolves
Great Ape Conservation Act (2000), 88
Great apes. *See* orangutans
Great Apes Survival Project (GRASP), 88
Great Lakes region, 41, 42
Green turtles, 57
Guedes, Neiva, 101–103
Gulf of Maine, 75

Habitat destruction and loss
  African elephants and, 51
  amphibian decline and, 10
  chimpanzees and, 53
  giant panda decline and, 50
  gorillas and, 54
  marine turtles and, 57
  tigers and, 55, 56
Hakalau Forest National Wildlife Refuge, 78, 83–84
Harrison, Dan, 46
Hawaii, 5, 37, 83–84
Health issues, xii
Home Depot, The, 93, 95
Howard, Bob, 81, 82

# INDEX


Human-related factors, bird extinction and, 17
Hunting. *See also* Poaching
  bushmeat, 108–111
  ivory trade and, 51–52
  of tigers, 55
Hyacinth Macaw Project, 99–101, 102–103

Iceland, whaling and, 22, 27
Idaho, 43
Illegal trade. *See* trade
India, 105
Indian Creek, Virginia, 36
Indian Ocean, 25–26
Indian rhinos, 48
Indigenous peoples
  bushmeat hunting and, 108–111
  protected areas and, 69–70
  turtle project in Indonesia and, 106
Individual actions. *See also* indigenous peoples
  illegal logging products and, 94, 95, 96
  influence of, xiii–xiv
  on national wildlife refuges, 80
  not using and buying products, 7–8
  preserving biodiversity and, 7
Indonesia, 92, 93, 94, 95, 96
Infectious diseases. *See* disease
International community. *See also* CITES (Convention on International Trade in Endangered Species)
  conservation efforts by, xxv–xxvi
  whale conservation and, 24
International Convention for the Regulation of Whaling (ICRW), 24, 28–29
International Monetary Fund (IMF), xxvi
International policy, 67–68
International Whaling Commission (IWC), xxv, 22–23
  in the 21st century, 28–29
  conservation agenda of, 24–25
  current situation in, 27
  establishment of, 24
  successes and failures of, 25–26
Invasive species, 11–12
Iowa, 84
Iridovirus, 11
Island biogeography, 3
Ivory ban, 52
Ivory trade, 51–52

Japan
  marine turtles and, 58
  whaling and, 22, 24, 26, 27
Jarvis, Ron, 93, 95
J. Clark Salyer National Wildlife Refuge, 81–82
Jeffrey, Jack, 84
Julie B. Hansen National Wildlife Refuge, 82–83

Kaong River (Cambodia), 103
Keystone species, 48
Khmer Rouge soldiers, 104, 105–106
Klamath-Siskiyou region, 45
Koko Foundation, 88
Kurth, Jim, 80
Kyoto Protocol, xii, xxi

Lake Erie, xxi
Lange, Bernadette, 102
Lantjoue (Cameroon), 108–111
Lassen County, California, 37
League of Nations, 24
Leatherback turtles, 57, 79–81
Leopold, Aldo, xxiii, xxvi
Lishman, William, 84
Locke, John, xxiii
Locust Fork, Alabama, 36
Logging
  bar codes to stop, 94, 95, 96
  chimpanzee decline and, 53
  efforts to combat illegal, 93
  hunting restrictions and, 109
Lossi Gorilla Sanctuary, 55
Lovejoy, Tom, 6
Lower Tennessee Cumberland ecosystem, 36

MacArthur, Robert H., 3
Macaw birds, 98–103
Magazine Mountain shagreen, 34–35
Maine, 46
Malakoff, David, xii
Malaysia, 105
Marine ecosystems
  management of, 73–74
  marine reserves and, 77
  threats to, 72
Marine Life Protection Act, 76
Marine protected areas (MPAs), 66, 72

Marine reserves, 72
  authority to create, 76–77
  ecosystems context of, 77
  location of, 75–76
  nature of, 74–75
Marine Resources Protection Act Ecological Reserve (California), 76
Marine sanctuaries, 72, 75
Marine Sanctuaries Act (1972), 72
Marine turtles, 48, 57–60
Marshall, Andrew J., 93, 94
Matthiessen, Peter, xi
Mauna Kea, 83
McCormick, Steve, 92, 94, 95
*Mesodon magazinenis* (Magazine Mountain shagreen), 34–35
Minnesota
  refuges in, 83
  wolf population in, 43, 47
Mississippi gopher frog, 38
Mladenoff, David, 46
Montreal Protocol, xxv–xxvi
Muir, John, xxiii, xvi–xvii

National Audubon Society, 18
National Conservation Programme (China), 49
National Marine Sanctuary Program, 76
National Oceanic and Atmospheric Administration (NOAA) Fisheries, 33
National Research Council, 33
National Wildlife Federation, xxiv
National Wildlife Refuge System, xiv
  in Alaska, xxiv
  challenges and limitations of, 82–83
  for desert bighorns, 82
  first refuge of, xix
  humans helping creatures for, 80
  landscape restoration by, 84
  for leatherback turtles, 79–81
  number of refuges in, 78
  restoring native ecosystems with, 83–84
  saving species with, 78–79
  for waterfowl, 81–82
  for whooping crane, 84–85
Native people. *See* indigenous peoples
Natural Resources Defense Council (NRDC), xxiv
Nature Conservancy, The, 3, 88, 89, 90, 91–92, 93, 94, 95, 96
Neal Smith Refuge, 84
Necedah National Wildlife Refuge, 85
Nevada, 37, 82
New Hampshire, 46
New Management Procedure (NMP), 25
Newts. *See* amphibians
New York, upstate, 46
Nickens, T. Edward, 79
North America, marine reserves in, 75–76
North American Breeding Bird Survey, 18
North Dakota, 81
Norway, whaling and, 22, 26, 27

Ocean habitats. *See* marine ecosystems
Ocean sanctuaries, 67
Odzala National Park, 55
Olive ridley turtles, 57–58
*On Human Nature* (Wilson), 3
Orangutans
  decline in, 89–91
  protection of, 90–92, 93, 94
  surveying population density of, 89–92
Organic products, xiii
Owl, 18–19
Ozark National Forest (Arkansas), 34

Palo, Harold, 98
Pandas. *See* giant panda
Pantanal, Brazil, 98–103
Parasitic chytrid fungus, 11
Pathogens, 11
Patuxent Wildlife Research Center, 85
Pelican Island, 78
Peregrine falcon, 33
Pesticides, xiii
Pew Oceans Commission, 72–73
Plan of Implementation of the World Summit on Sustainable Development, 67–68
Plicate rocksnail, 36
Poaching
  of marine turtles, 57–58
  of tigers, 56
Pollution, marine turtles and, 59
Pope, Carl, xiv
Prairie bush clover, 34
Predators, of marine turtles, 60

# INDEX


Protected areas. *See also* marine reserves; National Wildlife Refuge System
- benefits of, 63
- business practices and, 68–69
- climate change and, 67
- of ecoregions, 64
- effective management of, 63
- forests, 65–66
- freshwater, 66
- future work for, 70–71
- governance and, 68
- marine, 66, 72
- partnerships for, 64
- people and, 69–70
- policy development for, 67–68

PT Daisy, 96
PT Sumalindo Lestari Jaya, 96
Puget Sound, Washington, 75–76

Quaking aspen (tree), 44

Rabbit
- Columbia Basin pygmy, 38
- riparian brush, 35

Raccoons, 60
Ramsar sites, 66
Raven, Peter, 6
Reagan, Ronald, xxiv
Recycling, xiv
Refuges. *See* National Wildlife Refuge System
Refúgio Ecológio Caiman, 99
Regulation. *See also* CITES (Convention on International Trade in Endangered Species)
- state laws, xiii
- of whaling, 22–23, 24
- wolf recovery and, 41–42, 43

Reserves, giant panda, 49. *See also* marine reserves
Revised Management Scheme (RMS), 26
Revised Management Procedure (RMP), 26
Riparian brush rabbit, 35
Ripple, William, 44–45
Roosevelt, Theodore, xv, 78
Russia, 56

Sahelian plains, 14
Salamanders. *See* amphibians
Sanctuaries
- marine, 72, 75
- ocean, 67
- whale, 25–26

Sandy Point National Wildlife Refuge, 79–80, 81
Sawhill, John, 2
Schaller, George, 49
Sea otters, 72
Sea urchins, 72
Seideman, David, x
Semlitsch, Raymond D., 9, 10–12
Shade-grown coffee, xiv, 7–8
Sheep, bighorn, 82
Shrimping operations, 58
Sickley, Ted, 46
*Silent Spring* (Carson), xix, xxiii, 13
Sizer, Nigel, 95, 96
Snail, 34–35
Snape, William, 43
Sociobiology, 4
*Sociobiology* (Wilson), 4
Souris River, 81
South Africa, 76
South Carolina, 84
Southern Appalachian ecosystem, 36
Southern Ocean Whale Sanctuary, 26
Sovannara, Heng, 103, 104, 106, 107
Soviet Union, 26
Sre Ambel River region, 105, 107
Sri Lanka, 19
State laws, xiii
Stehn, Tom, 78–79, 85
Storks, 14–15
Sre Ambel River region (Cambodia), 105
Sumatra, 89, 90

Tan riffleshell, 36
Texas, 84
Texas City Prairie Preserve, 38
Thailand, 106
Thoreau, Henry David, xxiii
Tigers, 48, 55–57
Toads. *See* amphibians
Tortoise, gopher, 38
Tortugas Ecological Reserve, 76
Trade
- amphibian, 12
- bushmeat, 53, 54, 109
- giant panda, 50
- ivory, 51–52
- regulation of international, by CITES, 31
- of tiger parts, 55
- turtle products, 58

Traditional Chinese medicine, 56
"Trophic cascade," 44
Turtles
- batagur baska, 104–108
- marine, 57–60
- refuge for, 79–81

Ultraviolet (UV-B) radiation, 10
Umbrella species, 48
United Kingdom, 18

United States. *See also* National Wildlife Refuge System
bird decline in, 18
future of biodiversity in, 4–5
history of conservation in, xxii–xxiv
marine reserves in, 75–76
protection of endangered species in, 33–39
wolf population in, 40–47
U.S. Agency for International Development, 93, 95
U.S. Fish and Wildlife Service (FWS), 33, 41, 93
U.S. Geological Survey (USGS), 9, 85
Usongo, Leonard, 108, 110
U.S. Virgin Islands, 79

Vermont, 46
Virginia, 36

Washington
Columbia Basin pygmy rabbit in, 38
marine reserves in, 75–76
refuges in, 82–83
Washoe County, Nevada, 37
Wass, Richard, 83
Waterfowl habitat, 81–82
Water pollution, xii, xxi–xxii
Western lowland gorilla, 55
Western Sambo Ecological Research Reserve, 76
Wetlands, 10, 66
Whales
conservation of, 27–29
endangerment of, 22
regulatory group managing, 22–23
sanctuaries for, 25–26
Whaling
countries objecting to moratorium on, 27
history of, 23
key dates in, 28, 29
moratorium on, 22, 25, 26
regulation of, 24, 26
unregulated, xxv
White rhinos, 48
White storks, 14–15
Whooping crane, 84–85
Wildlife Conservation Society (WCS), 103, 104, 106
Wilson, E. O., x–xi
biodiversity and, xvi, 2, 3, 4–8
career of, 3–4
on conservation, xx
on shade-grown coffee, xiv
Wisconsin, 85
Wolves
biodiversity in Yellowstone National Park and, 44–46
legislation effecting recovery of, 41–42, 43
myths and persecutions of, 42–43
needed recovery programs for, 46–47
recovery of, 35, 41
Wood Buffalo National Park, 84
World Watch Institute, 13
World Wildlife Fund (WWF), xxiv, xxvi
bushmeat hunting and, 108–111
changes in, 6–7
on CITES, 30–32
conservation efforts rewarded by, 70
conservation strategies of, 64–69
endangered apes and, 88
on endangerment of flagship species, 49–60
global partnerships of, 63
on indigenous populations, 69–70
international collaboration and, 68
mission of, 48
whaling and, 22, 25

Yellowstone National Park, xxiii
wolf population and, 47
wolf recovery and, 44–46
Youth, Howard, 14

Zeller, Bruce, 82, 83

## ABOUT THE CONTRIBUTORS

YAEL CALHOUN is a graduate of Brown University and received her M.A. in Education and her M.S. in Natural Resources Science. Years of work as an environmental planner have provided her with much experience in environmental issues at the local, state, and federal levels. Currently she is writing books, teaching college, and living with her family at the foot of the Rocky Mountains in Utah.

Since 2001, DAVID SEIDEMAN has served as editor-in-chief of *Audubon* magazine, where he has worked as an editor since 1996. He has also covered the environment on staff as a reporter and editor for *Time*, *The New Republic*, and *National Wildlife*. He is the author of a prize-winning book, *Showdown at Opal Creek*, about the spotted owl conflict in the Northwest.